*Clinician's Guide to Interpersonal
Psychotherapy in Late Life*

Clinician's Guide to Interpersonal Psychotherapy in Late Life

Helping Cognitively Impaired or Depressed Elders and Their Caregivers

Mark D. Miller

OXFORD
UNIVERSITY PRESS

2009

OXFORD
UNIVERSITY PRESS

Oxford University Press, Inc., publishes works that further
Oxford University's objective of excellence
in research, scholarship, and education.

Oxford   New York
Auckland   Cape Town   Dar es Salaam   Hong Kong   Karachi
Kuala Lumpur   Madrid   Melbourne   Mexico City   Nairobi
New Delhi   Shanghai   Taipei   Toronto

With offices in
Argentina   Austria   Brazil   Chile   Czech Republic   France   Greece
Guatemala   Hungary   Italy   Japan   Poland   Portugal   Singapore
South Korea   Switzerland   Thailand   Turkey   Ukraine   Vietnam

Copyright © 2009 by Oxford University Press, Inc.

Published by Oxford University Press, Inc.
198 Madison Avenue, New York, New York 10016

www.oup.com

Oxford is a registered trademark of Oxford University Press

All rights reserved. No part of this publication may be reproduced,
stored in a retrieval system, or transmitted, in any form or by any means,
electronic, mechanical, photocopying, recording, or otherwise,
without the prior permission of Oxford University Press.

Miller, Mark D., 1955–
Clinician's guide to interpersonal psychotherapy in late life : helping cognitively impaired
or depressed elders and their caregivers / Mark D. Miller.
    p. cm.
Includes bibliographical references and index.
ISBN 978-0-19-538224-2
1. Depression in old age.   2. Psychotherapy.   3. Older people—Care.   I. Title.
RC537.5.M538 2009
618.97'68527—dc22
                                                                2008043412

9 8 7 6 5 4 3 2 1

Printed in the United States of America
on acid-free paper

Preface

This manual will outline the conceptual rationale and practical application of a variation of interpersonal psychotherapy (IPT) for use with cognitively impaired older adults (thus renamed IPT-ci). IPT was developed as an evidence-based treatment for depression; however, clinical experience indicates it is also beneficial for older adults with cognitive impairment even without coexisting depression.

The author's background experience is based upon work as a geriatric psychiatrist for 21 years in a university-based multidisciplinary outpatient geriatric treatment center (Benedum Geriatric Center, University of Pittsburgh Medical Center) as well as work with his contributing colleagues to refine the modifications of IPT for use with cognitively impaired elders through experience gained from implementing IPT in three randomized trials with depressed elders: MTLD-1 (Reynolds et al., 1999), MTLD-2 (Reynolds et al., 2006), and IPT for Grief (Shear, Frank, Houck, & Reynolds, 2005). Using IPT with older individuals in these research protocols required very little modification of the original techniques developed by Gerald Klerman, Myrna Weissman, and colleagues. This manual of guidelines builds upon the basics of IPT by adding extra tools and techniques specific to working with a cognitively impaired population.

To provide some key background material for students of IPT who are new to the field of geriatrics, this manual is divided into two sections. In Section I, the introductory chapters provide an overview of traditional IPT principles, including illustrative case vignettes of geriatric patients. Additional concise chapters on gerontology/geriatric medicine, late-life depression, cognitive impairment/dementia, and executive dysfunction are also included in order to equip the reader (particularly the beginning student) with an adequate point of reference to recognize the typical kinds of problems and issues that arise in working with this population.

Section II is devoted to the actual application of IPT-ci in detail with ample case vignettes (there are 32 case vignettes total throughout the book; case vignettes are real patient scenarios with changes made to protect confidentiality) to illustrate the approach to the initial interview, the integration of concerned family members or caregivers from the outset of treatment, the impact of cognitive impairment on the four IPT problem areas (unresolved grief, role transition, role dispute, and interpersonal deficit/sensitivity), the caregiver's own role transition, and strategies for reaching a steady state of maximized coping or functioning. IPT-ci also incorporates long-term maintenance strategies, and this manual will demonstrate how the IPT-ci therapist can continuously help the identified patient and his or her caregivers to adjust to new problems, changing circumstances, or further declines in cognitive impairment with the goal of continuing to optimize functioning for the remainder of the identified patient's lifetime.

IPT-ci is a workable strategy for approaching the total long-term psychosocial care needs of the older patient for social workers, psychologists, nurse practitioners, or physicians. As IPT-ci builds upon the principles of traditional IPT, the added value of the IPT-ci strategies will equip the therapist to cover the gamut of presenting problems in older adults—from depression in cognitively intact individuals to those who show signs of early cognitive changes or minimal cognitive impairment (MCI), particularly those who show executive dysfunction, which is often unrecognized or misunderstood by caregivers. Finally, the IPT-ci format can also continue to help caregivers and patients with frank dementia to cope maximally.

The goals of IPT-ci are to provide ample psychoeducation for patient and caregivers, to complete an interpersonal inventory, to encourage an adequate biomedical evaluation to search for treatable causes of depression or cognitive impairment, to use psychotropic medication concomitantly as indicated, to engage the patient in the appropriate IPT problem area, to utilize single or joint sessions with the patient, caregiver, or both, and to optimize the IPT treatment strategy to work to resolve role disputes. Ultimately, the long-term goal of IPT-ci is to work to achieve a steady state of optimized education, depressive symptom resolution, and functionality that is appropriate for the patient's current cognitive status. IPT-ci provides a strategy to continue to work with patients whose cognitive status evolves into a diagnosable dementia whether or not depression coexists. Special attention is given to caregivers or concerned family members who are going through their own role transition in becoming caregivers by integrating them into the entire treatment process from the first meeting.

Acknowledgments

The late-life depression research group at the Western Psychiatric Institute and Clinic (WPIC) (University of Pittsburgh) has utilized IPT in various depression research protocols, including several studies in geriatric-aged patients under the particular leadership of Charles F. Reynolds III, in collaboration with Ellen Frank, David Kupfer, Kathy Shear, Cleon Cornes, and the late Stanley Imber. The collective experience of the therapists who worked in these research projects and collaborators from related fields helped to refine the concepts of IPT-ci, which continue to evolve. The contributions of a variety of WPIC colleagues in their role either as IPT therapists, research collaborators, or Benedum Geriatric Center colleagues deserve to be acknowledged: Allan Zuckoff, Jennifer Morse, Lynn Martire, Lin Ehrenpreis, Lee Wolfson, the late Rebecca Silberman, Valerie Richards, Scott Kaper, Julie Marx, Barbara Tracey, Elizabeth Elliot, Jackie Stack, Elizabeth Weber, Mike Lockovich, Jill Houle, Kristin Carreira, Julie Malloy, Jean Zaltman, Mary Ganguli, Rick Morycz, Diane Conti, Ann Zachariah, Carol Leach, Joe DeSantis, Shirley Mucha, and Richard Schulz. Special thanks to Eric Miller, Diana Donnelly, Mary Herschk, and Julia TerMaat for their editorial assistance.

Contents

Section I Rationale and Background for IPT-ci

Introduction 3

1. Overview of Basic Principles of Interpersonal Psychotherapy 8
2. Overview and Background for IPT-ci 21
3. A Concise Review of Gerontology/Geriatric Medicine 30
4. Overview of Late-Life Depression 37
5. The Cognitive Impairment Spectrum: MCI to Dementia 50
6. Manifestations of Executive Dysfunction 64

Section II IPT for Cognitive Impairment (IPT-ci)

7. IPT-ci Basics 87
8. Incorporating Family/Caregivers Into the Treatment Process From the First Meeting 103
9. Specific IPT Foci in IPT-ci: Grief, Role Transition, Role Dispute, Interpersonal Deficits/Sensitivity 117
10. The Caregiver's Own Role Transition 147
11. Flexible Individual and Joint Sessions 158
12. Reaching Steady State and Long-Term Planning 163

Postscript Future Directions for IPT-ci 175

Appendix Montreal Cognitive Assessment (MoCA) Administration and Scoring Instructions 179

References 183

Index 191

Section I

Rationale and Background for IPT-ci

Introduction

As the demographers predict an ever-increasing proportion of elderly citizens in our population, particularly as the baby boomers in the United Sates turn 60, the health care demands of caring for this group are predicated to be enormous. We are now surviving longer by successfully fighting traditional killer diseases such as heart attacks, strokes, and cancer, with larger numbers enjoying the highest quality of life of any time in history. The downside of this equation is that more of us will become demented from Alzheimer's disease, which typically begins to be noticed in the seventh decade of life with a prodromal period of milder symptoms sometimes referred to as minimal cognitive impairment (MCI). Although we know a great deal about the various causes of cognitive decline, we do not know how to arrest the relentless march of Alzheimer's disease, the most prevalent cause of dementia by far. Current biomedical interventions slow the course of the disease at best.

On the severe end of the dementia spectrum, innovations in care such as Alzheimer Units and day programs are in place and growing in number to provide a safe supportive environment and to try to maintain as much dignity as possible for the afflicted individual. Less work has focused on the earlier phases of cognitive decline although researchers are now trying to discern clues and make predictions as early as possible amid hopes that biomedical treatments will have the greatest benefit if applied early in the disease process. Our best indication today is that those individuals diagnosed with MCI who have prominent deficits in memory convert to a diagnosis of Alzheimer's dementia at a rate of about 15% per year (Petersen & Negash, 2008). As the biomedical research progresses, however, afflicted individuals need to be managed now with more than a proper diagnosis and the best putative biomedical intervention. Cognitive impairment, even in its early stages, can have profound consequences on quality of life, risks for depression, impaired social functioning, worsened interpersonal relations (leading to a declining support system), and even safety issues such as misjudging traffic risks, accidental self-injury, vulnerability for financial fraud, and even an increased risk for suicide.

Purpose of This Manual

The purpose of this manual of guidelines is to educate the health professional about the pragmatic strategies used in IPT-ci for interviewing, engaging, educating, and treating individuals with cognitive impairment with or without coexisting depression as well as their caregivers. For older individuals with depression who are cognitively intact, traditional IPT may be wholly adequate as a one-to-one weekly psychotherapy to successfully reduce their depressive symptoms. For those who show cognitive impairment as well as depression, the added techniques of IPT-ci will engage and incorporate caregivers in the therapeutic process. This manual prepares the student IPT-ci therapist to see the needs of the identified patient as inextricably linked to caregivers such that a comprehensive plan to work with both simultaneously has the best chance for a successful outcome. For the reader who is just being introduced to IPT, the overview of IPT in Chapter 1 is intended to be an adequate point of reference for understanding the modifications of IPT-ci.

IPT-ci is a strategic framework for working with cognitively impaired individuals (with or without depression) and their caregivers that is particularly useful in the early stages of cognitive impairment when the most misunderstanding exists, when subtle role transitions are beginning to take place in both the identified patient and the caregiver(s), and when role disputes commonly arise out of these misunderstandings.

Research Using IPT in Older Individuals

Although interpersonal psychotherapy (IPT) was originally designed as a treatment for reducing depressive symptoms in adults, it has since been modified to broaden its use for children and adolescents, for women who experience depression during pregnancy or the postpartum period, for the medically ill such as AIDS patients, and even for patients without depression who suffer from anxiety disorders (Weissman, Markowitz, & Klerman, 2007).

Over the past 20 years of our research team's work with depressed elders within various National Institute for the Mentally Handicapped (NIMH)-sponsored research protocols, we found IPT to be a user-friendly and helpful evidence-based treatment for depression in nondemented depressed individuals. Using combined treatment of antidepressant medication and IPT, we were able to achieve a remission rate of 78% in elders with recurrent major depression (Reynolds et al., 1999). Using IPT in combination with placebo pill showed statistical superiority to clinical management plus placebo pill as a maintenance treatment for preventing a recurrence of major depression. In order to obtain a "clean sample" of depressed elders not complicated by dementing disorders, in the Maintenance Therapies in Late-Life Depression Study (MTLD-1) (Reynolds et al., 1999), we excluded those with a Mini–Mental Status Exam (MMSE) score less than 27.

In subsequent studies, we broadened our inclusion criteria to more closely match a true cross section of elders in the community by including subjects with MMSE scores as low as 18/30 (MTLD-2) (Reynolds et al., 2006). This change meant that we began to see comorbid cognitive impairment in addition to depression in our patient sample. In our research protocols that called for a comparison of antidepressant medication to IPT or their combination, we worried about the fidelity of psychotherapy in subjects who showed memory impairment or limited insight due to cognitive impairment. Simultaneously, we noted that family members more often accompanied the subjects to their appointments when more cognitive impairment was obvious and were asking for advice about management issues or problems they were encountering. This was partly due to the older sample we were recruiting but we also noted that these family members were much more affected themselves and worried about their afflicted loved ones when cognitive impairment was present in addition to depression. It dawned on us that these concerned family members were in the process of becoming caregivers themselves! Some caregivers frequently accompanied subjects to every appointment and were instrumental in guaranteeing medication compliance and overall fidelity to the treatment, not to mention help with instrumental activities of daily living (IADLs), compliance with good medical care, good nutrition, and so on. Other caregivers were bewildered, angry, baffled, demanding, and poorly cognizant of the meaning and impact of the changes taking place before their eyes in their loved one. These cognitive changes were often, unfortunately, permanent and progressive in most cases. Needless to say, engaging the caregivers became essential for compliance and fidelity to the requirements of the research but it also became very clear that the caregivers' experience, level of understanding, and distress level also needed to be addressed even though they were not the "patient." It subsequently became clear that successfully addressing caregivers' concerns and assisting them to solve problems related to their afflicted loved one (and sometimes to help resolve interpersonal role disputes between them) made them better caregivers which, in turn, showed an obvious beneficial effect on the identified patient (the subject of the research) regarding his or her compliance with treatment, the completeness of his or her depression resolution, and his or her overall improved welfare and quality of life. It was these two issues (the concern about how an individual with cognitive impairment could fully utilize psychotherapy and the glaring necessity to engage, educate, and guide caregivers) that led to the modifications of IPT outlined in this manual.

Using IPT With Depressed Elders in a Research Setting

In planning to use IPT in a study designed to determine which treatment worked best as a maintenance treatment of depressed elders who were first treated to remission with combined treatment with paroxetine and IPT by design, the MTLD-2 study (Reynolds et al., 2006) recruited a mixed sample with some subjects having cognitive impairment and some who were cognitively normal (the MTLD-II Study). Subjects who remitted with combined therapy and remained in remission for 16 weeks were randomly assigned to one of four cells: antidepressant medication (paroxetine) plus IPT, antidepressant medication plus clinical management (CM), placebo pill plus IPT, and placebo pill plus CM.

The CM condition consisted of a review of symptoms but was deliberately not a therapy session and audio tapes confirmed this by independent raters. The drug and placebo tablets were identical and thus subjects and their therapists were blind to their drug status. Research clinicians who obtained the independent mood ratings from the patients were also blind to drug assignment and to IPT versus CM assignment. The results showed that those who received antidepressant medication fared the best in terms of the lowest recurrence rates for major depression. The IPT plus placebo group did not separate from the CM plus placebo group, which was surprising since this comparison did show statistical superiority in the prior mentioned study of younger aged geriatric subjects (mean age 68), who were required to be cognitively intact (MTLD-1). Several of the same IPT therapists participated in both studies. These puzzling findings prompted secondary analyses of the data, in which we found that those subjects with cognitive impairment who received maintenance IPT plus placebo pills actually did show better long-term survival without a recurrence of their depression that was statistically superior to the cognitively impaired subgroup who received CM and placebo (Carreira et al., 2008). This finding also surprised us as we expected the opposite—that the more cognitively impaired subjects would benefit less than cognitively normal subjects. How could we explain why IPT seemed to offer greater protection against a recurrence of depression than CM in the cognitively impaired group but not in the cognitively intact group?

One hypothesis for why the cognitively intact group receiving IPT plus placebo did not separate statistically from the same group receiving CM plus placebo is type II error explained by the CM condition being inadvertently more like supportive therapy than intended, thus making it more difficult to detect a difference between groups. The reason we believe this may be so is that when we decided to do this study with depressed elders, some of whom would also have cognitive impairment, we (as a team) acknowledged our misgivings about trying to successfully implement IPT with the cognitively impaired patients whom we expected to have less recall ability and perhaps less capacity for insight and perhaps less capacity for change. We reasoned that in order to attempt to work with this group, we would have to utilize the family members who often accompanied the subjects (more often than in our earlier MTLD-1 study) to discern or confirm the identified patient's true functioning from week to week. We further reasoned that we would need to rely on family members more so to implement changes in behavior in order to help the identified patient to recover from depression. For example, if John is a depressed subject with memory loss, and he and his IPT therapist were working on ways in which he could reengage with some of his friends he had stopped visiting in order to become more active (and therefore feel less isolated and lonely), instead of relying entirely on John's memory to carry out that which he had agreed to during the IPT session, we made sure to mention the plan to the accompanying spouse or adult child who then reminded the subject and often facilitated carrying out the plan.

The cordial demeanor of the IPT therapist was also appreciated by many subjects who said they liked their meetings because they felt listened to and cared about in a nonjudgmental way. We also noted that even though subjects with memory loss often could not recall specific details from their previous IPT sessions, with minimal prompting, they were able to get "back to the same page" and renew the discourse as if the meaning or importance of the session was accessible even if the details were not spontaneously recalled (these subjects had cognitive function in the mild to moderately impaired range). Perhaps the change to less specifically integrated interventions between patient and caregivers in those cognitively impaired subjects assigned to CM and placebo accounts for the finding of statistical superiority for the IPT and placebo group in preventing a recurrence in the maintenance phase.

Research ideally comparing IPT-ci to a time-matched control that utilizes outcome measures of depression severity, functional capacity, and quality of life in the identified patient and depression severity and caregiver burden-induced stress in the caregiver needs to be completed for further evidence-based exploration of the validity of IPT-ci.

The next chapter will deepen the discussion of the rationale and background for IPT-ci.

1
Overview of Basic Principles of Interpersonal Psychotherapy

Interpersonal psychotherapy (IPT) is a manual-based psychotherapy originally developed for the acute treatment of depression. IPT can be used alone or in conjunction with antidepressant medication. IPT focuses on interpersonal relationships in the patient's current life that

1. may have contributed to the depressive symptoms (such as chronic marital discord or friction with an overly critical boss);
2. may have been damaged by the depression (someone who has repeated depressions from more biological roots; e.g., an individual with a strong family history of depression might experience a drop in energy level, lack of motivation, oversleeping, and a decrease of interest in sex with no obvious precipitant, which, in turn, might make his or her spouse feel lonely and neglected); and
3. the patient can derive more support from to help him or her recover from depression (e.g., a shy person who feels depressed due to lack of friends or social outlets might be encouraged to discuss ways to nurture a new friendship to improve his or her flagging self-esteem, thus reducing his or her depressive symptoms).

Development of IPT

IPT was developed by a collaborative effort by Klerman, Weissman, Rounsaville, and Chevron (1984); they compiled empirical data from social psychology, attachment theory, and interpersonal psychology, which they refined during their work with depressed patients in a depression treatment clinic. These collaborators published their original manual of guidelines in 1984 and a revision in 2000 (Weissman, 1999). These researchers noted that although the syndrome of major depression can be predisposed by genetic or biological, environmental or developmental factors, most problem areas that preceded the onset of the depressive symptoms have something to do with current interpersonal relationships and fall into four broad categories: unresolved grief, role transition, role dispute, or interpersonal deficit (sometimes called interpersonal sensitivity). Table 1.1 summarizes the 3 components of IPT.

Table 1.1 IPT Views Depression as Having Three Parts

1. Depressive symptoms as outlined in the *DSM-IV-TR* manual (American Psychiatric Association, 2000)
2. Social and interpersonal factors. How satisfying are key relationships in personal or family life, work, religious affiliations, friends, etc.? What social stressors might be contributing to a risk for depression?
3. Personality factors—these are enduring traits that are formed during childhood that constitute a person's character or personality. Personality traits can be categorized along many dimensions such as outgoing or shy. Maladaptive traits (such as low frustration tolerance or overdependence) can be risk factors for depression. Personality traits that cluster together to form a pattern of maladaptive behavior are known as personality disorders. The goal of IPT is to reduce depressive symptoms and to bring about adaptive change by focusing on interpersonal relationships and changes in social roles. Although IPT recognizes the contribution of personality traits, it does not attempt to change personality as that would be unrealistic in the 12–16 week duration of a typical course of IPT.

The interested reader is referred to an excellent brief synopsis of the tenets of IPT for the practicing clinician entitled *Clinician's Quick Guide to Interpersonal Psychotherapy* (Weissman, Markowitz, & Klerman, 2007). See the list of references for further resources focusing on older patients (Post, Miller, & Schulberg, 2008; Schulberg et al., 2007; Stuart & Robertson, 2003).

Phases of IPT

The format of IPT can be divided into four phases.

Phase I

Phase 1 encompasses steps that engage the patient in an empathetic way to encourage a supportive alliance for working together successfully. The following are the steps in Phase I:

1. Complete a psychiatric evaluation and differential diagnosis.
2. Obtain a thorough medical evaluation to rule out potential medical contributions to depression.
3. Provide ample psychoeducation about the biopsychosocial model of depression.
4. Contract for 12–16 weekly individual face-to-face sessions.
5. Give the patient the "sick" role to relieve his or her guilt for lack of social motivation.

6. Establish one of four foci or problem areas to concentrate on for treatment (unresolved grief, role transition, role dispute, or interpersonal deficit).
7. Explore in some depth all pertinent interpersonal relationships that affect the patient's current life (the interpersonal inventory).
8. Initial sessions focus on exploring the relationships identified in the interpersonal inventory and pointing out connections between changes in social functioning and the onset and maintenance of depressive symptoms.

Phase II

Phase II involves the following:

9. Continue to work on the established focus by helping patients to clarify their problems. Encourage them to admit their full range of feelings and affects and to link any decrease in depressive symptoms to changes made in social roles or the way they learned to handle problem relationships differently. Carry out communication analysis in role disputes and test perception and performance between sessions. Use decision analysis, and sometimes role play, to achieve better clarification of new and potentially better coping strategies.
10. Specific techniques are provided for each of the four foci. For example, in the role transition focus, the IPT therapist helps the patient mourn the loss of the old role and subsequently explore options for new satisfying opportunities.

Phase III

Steps in Phase III include the following:

11. Final sessions focus on consolidating gains and readying for termination by explicitly acknowledging that termination is a potential loss for the patient.
12. Terminate as per the contract within the allotted time frame under the original agreement (12–16 weekly sessions).
13. If good resolution of depressive symptoms is not achieved, consider a different type of psychotherapy or medication trial. If patients feel they need more therapy beyond the agreed-upon number of sessions, insist on at least a 3-month period of self-assessment before pursuing additional treatment to see if they can consolidate what they have learned on their own.

Phase IV

For patients with a history of recurrent episodes of depression, consider monthly maintenance therapy, particularly for those with a role dispute focus.

Structure of Therapy

The role of the IPT therapist is one of a trusted professional who plays an active role in explaining the facts about depression and obtaining the necessary information to make an accurate assessment and treatment plan. The IPT therapist is an educator and advocate but not a friend or social contact. The IPT therapist does not focus on early life events, transference, dreams, cognitions (as in cognitive-behavioral therapy [CBT]), repressed memories, or other intrapsychic themes. The goal of IPT is to advocate for change in social roles that reduces depressive symptoms, which is different from supportive therapy, in which the goal is often to learn ways to cope with or better accept stressful circumstances.

The patient is expected to work in IPT in order to improve. The patient's role is to attend the sessions regularly, think about the discussions between sessions, actively participate in problem solving, and try out alternative coping strategies for improving social relationships between sessions.

The contract for a limited number of sessions (12–16) places conscious and unconscious pressure upon the patient to work effectively in IPT before the allotted time runs out. New events that come up during the therapy are acknowledged but patients are gently steered back to the agreed-upon focus in order to make progress in reducing depressive symptoms within the short-term time frame.

Although IPT was designed as an individual treatment, variations have been modified for group settings and for couple's therapy. In individual therapy, it is sometimes advisable to meet the spouse or other significant person to clarify what the therapy process is and isn't and to hear his or her alternative perspective; however, the subject matter and the goals for such a meeting need to be explicitly spelled out prior to the meeting such that confidentiality and the individual-based treatment strategy are not compromised.

Benefits of IPT

Some of the selling points of traditional IPT for patients, in addition to its deceptive simplicity, are as follows:

1. the formulation of a treatment strategy that focuses on one or two problem areas in one of four broad categories (unresolved grief, role transition, role dispute, or interpersonal sensitivity/deficit);
2. its comprehensive assessment of all the important people in the afflicted person's life (the interpersonal inventory) and the utilization of that information to explore alternative coping strategies involving one or more of the identified important relationships;
3. the subtle and not-so-subtle pressure of a time-limited contract to stay on track toward making progress in exploring changes in social roles that lessen depressive symptoms; and

4. the designated short-term treatment contract (3 to 4 months) allows for successful termination without engendering cumbersome transference reactions that could potentially lengthen the termination process.

Proponents of IPT appreciate the conversational style and the extensive psycho-educational component in which the IPT therapist is a facilitator and a teacher who encourages his or her patient to seek greater understanding and to explore alternative coping strategies to improve functioning in social roles and thus reduce depressive symptoms. These interactions are often perceived by patients as a seamless segue from a conversation with an empathetic, trained professional who advocates for their best interest. Patients have appreciated the use of terms and concepts they can easily grasp even if they have no prior experience in therapy. Patients who lack keen psychological mindedness or a formal education are still able to benefit from IPT.

Another strength of IPT is its empirical underpinnings and the extent to which it can be taught to a variety of healthcare personnel without requiring years of training. With 20 hr of didactic instruction (one intensive weekend) and 2–3 completed cases under qualified supervision, any motivated physician, psychologist, nurse, or social worker can become proficient in IPT. In the United States, the majority of psychotherapy is, in fact, delivered by social workers in the broadest sense.

Comparison to Other Treatments

The most frequently cited evidence-based psychotherapy is CBT and its derivatives such as problem-solving therapy (PST). Clearly, CBT/PST has its own literature demonstrating effectiveness and a dedicated following of practitioners (Alexopoulos, Raue, & Arean, 2003).

With its focus on cognitions and specific behaviors, rather than on social roles or feeling states, CBT/PST is perceived by some patients as too burdensome in terms of homework assignments and to-do lists such that some abandon it early on. Other patients are less interested in self-exploration and prefer the more structured approach of CBT. The developers of IPT acknowledge that it is not a panacea and that some patients may prefer or do better with CBT or another form of psychotherapy (or medication trial or electroconvulsive therapy), particularly if significant progress is not made after a 12–16 week trial of IPT.

Use of IPT With Cognitively Intact Geriatric Patients

Very little adaption of IPT is required for geriatric patients who are cognitively intact (we will discuss cognitive impairment in Chapters 5 and 6). Older patients may need accommodation for hearing loss, may have transportation difficulties, or may be unable to tolerate a full hour-long session due to medical illness or chronic pain. Telephonic sessions may be necessary if attendance in person cannot be weekly. Older patients have shown themselves as a group to be as capable as younger patients to benefit from IPT. In fact, some advantages of working with older patients include flexible schedules and a tendency to be less inhibited to speak more freely and openly than younger patients. Hinrichsen and Clougherty (2006) provide an excellent book-length review of IPT with cognitively intact geriatric patients and a summary of the research behind IPT.

The following case vignettes will illustrate the use of IPT with cognitively intact geriatric patients.

CASE VIGNETTE 1: COMPLICATED GRIEF

Patricia, age 61, was referred for IPT after her husband died by suicide. She did not see it coming at all and throughout the joint work in IPT, she and her IPT therapist pieced together a likely motive related to her husband being recently passed over for a promotion, in addition to his longstanding unhappiness in his career. One technique utilized in IPT for grief is to spend considerable time reviewing the details right before and after the death. In further discussion with family and from what she had observed throughout their 25 years of marriage, it became clear to Patricia that her husband was a great internalizer of his feelings and had continued on a high-pressure career path even though he no longer liked it and felt trapped by his obligations. His suicide note to her, which said that he loved her but that he could no longer stand the pain of his life, indicated that he was probably depressed but too proud to seek help or reveal the extent of his pain even to her. Patricia struggled with feeling enraged toward him for not giving her a chance to try to help him alternating with sadness at the level of his desperation.

Patricia's IPT therapist helped her by encouraging her to express all of her emotions both positive and negative and by giving her a sense that what she was feeling was within the normal range of others who had gone through similar traumatic experiences. After 14 weeks of regular meetings, she finally concluded that she must not have known her husband as well as she thought she had and came to the conclusion that that her husband was an "emotional island" in many ways when she began to chronicle a series of events and interactions over the years of their marriage. She also expressed relief that they had chosen not to have children because it spared them the pain that she now felt. After a full 16-session course of IPT, Patricia felt that she was growing weary of "rehashing" so many details of her husband's life and untimely death. She agreed to terminate as per the contracted number of sessions and 6 months later sent a thank you card indicating that she had started dating.

Case Vignette 1: Discussion

The goal for IPT with a focus on grief is to create a comfortable environment where the patient can begin or continue the grieving process that has been distorted or thwarted for some reason. In Patricia's case, her anger and rage over "not being consulted" before her husband took his life made her feel so guilty that she tried to forbid herself from thinking about him, which led to her feeling hopelessly stuck and depressed. The persistence, calmness, matter-of-factness, and gentle encouragement provided by her IPT therapist seeking clarification and the expression of affect allowed the patient to reenter the emotions of grief that had been suspended until her anger toward her late husband could be addressed. Behind the anger was a flood of sadness and dismay that both of their lives had turned out as they did. After ample opportunity was provided for exploration of speculated motivation on his part and her reaction to his suicide and the expression of strong emotions of anger and sadness to come forth, she felt considerably relieved and less depressed. She concluded that she might never know his true motivations to take his own life but that she had to move on regardless and eventually began to date other men.

CASE VIGNETTE 2: ROLE TRANSITION—ADJUSTING TO RETIREMENT

Joe, a welder in a steel mill, was forced to retire at a mandatory age. He had loved his job, particularly the creative requirements of finding solutions to maintenance problems in the mill. During the winter months of his first year of retirement, he became depressed and spent most of his day sitting alone in his basement in front of a darkened television as if in a trance. He would make statements that indicated that he no longer felt he had a purpose since his children were grown and he was "not needed" at the mill. All that was left, in his view, was to wait to die, he once said to his wife. His daughter finally convinced him to go for help and he reluctantly agreed.

The focus of IPT, in Joe's case, was clearly role transition from feeling vital as a sought-after expert welder to feeling purposeless and unneeded in retirement. His IPT therapist helped him to mourn the loss of his old role by encouraging him to reminisce about all of his successes and his pride in the steel that went into landmark buildings he had helped to create. In subsequent sessions, in attempts to help him to explore possible new roles he might play that would provide substitute satisfaction, it became clear that he had not cultivated any hobbies except his devotion to the Pittsburgh Steelers football team, but their season was now over for the year.

He liked to tinker and fix things but had already run through the backlog of projects he had intended to complete upon retiring. A serendipitous event then took place during Joe's course of IPT. His wife insisted that they take advantage of an opening in a retirement community she had been keeping her eye on for 2 years. Though Joe said that he merely "went along with the idea," with his IPT therapist having primed him to be "on the lookout for opportunities" to utilize his skills and seek new satisfactions, he busied himself with fix-it projects in his new environment. As they were gradually introduced to their new neighbors in the retirement village, word got around to several widowers in the community that Joe was handy with tools and he began to receive requests to hang doors and fix leaking faucets. When recounting these activities, Joe seemed to beam with pride. His mood improved greatly and his IPT sessions came to an end.

Case Vignette 2: Discussion

In Joe's case, there was little need for collaboration with family as both his referring daughter and his wife showed a good grasp of why Joe was stuck in his depression. They were unwaveringly supportive and encouraging to him without any need for clarification from his IPT therapist.

In contrast to the case of Joe, in which traditional IPT was wholly adequate to embrace the role transition focus that was driving the episode of Joe's depression, in the next vignette, adjusting to retirement was the logical starting point but an underlying role dispute eventually became the main focus of treatment. This scenario, where a role transition or grief focus eventually reveals an underlying role dispute that is more complicated, is not an unusual one.

CASE VIGNETTE 3: ROLE DISPUTE DURING ADJUSTMENT TO RETIREMENT

Germaine retired after working as a nurse for 35 years and looked forward to a more relaxed lifestyle in which she no longer needed to rise at dawn every day to go to work. She fell into a state of depression, however, as the life she eagerly anticipated seemed to fill up quickly with other responsibilities, among them regularly babysitting her grandchild. She had fantasized about volunteering in the local library, but she never seemed to find the time to inquire about how to go about signing up as a volunteer. Germaine felt guilty for feeling selfish and did not say anything to anyone about her sense of disappointment about how her retirement was turning out. During the first encounter with her IPT

CASE VIGNETTE 3 (*Continued*)

therapist, she stared down at the tissue in her hand and said she didn't deserve to take up time that could be better spent on someone else.

Her IPT therapist initially convinced Germaine that it was appropriate for her to assume the sick role as her depression had made it all the more hard for her to carry out her long list of self-appointed duties. The initial focus appeared to be the role transition to retirement. Her IPT therapist helped her to explore the barriers that kept her from asking for or pursuing what she really wanted and why she always seemed to defer her needs to the needs of others (perhaps this was one reason she became a nurse in the first place). After several sessions of communication analysis and role playing, Germaine reluctantly agreed to have a conversation with her daughter about her true desire to babysit for 2 days per week instead of 5. To Germaine's surprise, her daughter was very understanding and asked her why she had not said anything earlier. Her daughter contracted for a paid babysitter for the other 3 days per work week, thus allowing Germaine the freedom to finally be able to volunteer once a week at the local library.

After this role transition was "resolved," the focus of the therapy changed to one of role dispute. Her difficulties in her relationship with her husband had been alluded to briefly but not dwelled upon since the interpersonal inventory was completed. Now that the babysitting challenge was resolved, she sheepishly began to mention that she felt oppressed by her husband for a long time, describing herself as a "doormat" who never tried to stand up to him for several years. She described him as dictatorial and short-tempered historically. Her husband had become disabled in an industrial accident 12 years earlier and spent his day with a structured routine of walks and exercises to try to compensate for his physical disabilities. She felt oppressed because she perceived that she was expected to join him on his daily walks and exercise routines now that she was retired. She went on to describe how she felt that she had devoted her whole life to other people as a nurse, a mother, and a wife and could not look forward to any respite from this continuous stream of duties she felt expected to perform, which left her feeling depressed and trapped. After the initial success with the babysitting renegotiation, Germaine agreed to switch to a role dispute focus and also agreed to explore why she continued to take the stance she did with her husband and to ask herself how she might try to negotiate differently. When asked what she really preferred, she replied that she did not mind walking with him sometimes but she feared being locked in by the precedent.

The idea of confronting her husband with any of her own requests filled her with fear as she reported he had a temper and she did not want to be "shot down" or reprimanded. After some role play and rehearsal, she timidly agreed to try talking to him about it and managed to do so successfully. As with her daughter, her husband's response was surprising to Germaine. He was more than agreeable to have her come and walk with him whenever she chose to, but he said she was under no obligation to do so as he was used to going by himself for the past several years (after he had become disabled) while Germaine was still working. The responsiveness of both her husband and her daughter improved her self-esteem and lightened her depression as she gained more confidence that she could express her needs more clearly and directly and that her worst fears of being reprimanded for every request were overblown.

Case Vignette 3: Discussion

In discussing the case of Germaine, it is clear that she maintains a passive stance in her personality that landed her in many subordinate roles during her lifetime. Although changing personality is not a goal of IPT, the extent to which Germaine felt trapped, helpless, hopeless, and depressed was explored with clarification, solicitation of affect, and a willingness to take chances at negotiating differently. The crux of the issue that led to her feeling stuck and hopeless was her timidity in asking for what she wanted. This longstanding pattern in interpersonal relationships was clarified as the major factor in the maintenance of her depression and when asked if she wanted to work on ways to do it differently, she agreed, with some trepidation, to try. With the guidance, patience, and support of her IPT therapist, she was able to map out a better plan for her retirement that was closer to the one she had fantasized about and she was able to confront her husband who had a reputation for having a temper. She stood her ground successfully, which further bolstered her confidence that she could repeat the process in new situations that might arise in the future. The link between the reduction in depressive symptoms and the changes she was making in her social roles was specifically pointed out. By the end of her course of IPT, Germaine was no longer depressed and felt more adept at plumbing her own feelings before jumping to the aid of others who seemed to be in need of some help. She vowed not to jump in to help unless she felt she was ready to do so and the task was something that fit into her overall view of her retired life.

CASE VIGNETTE 4: INTERPERSONAL DEFICIT/SENSITIVITY— A CASE OF DEPENDENT PERSONALITY DISORDER

Millie was 61 years old when her mother, with whom she lived her entire life, died of a sudden illness. She was brought for treatment six months later after

CASE VIGNETTE 4 (*Continued*)

becoming depressed. Millie's version of the story was that she had taken care of her mother, and her father before that, out of her sense of duty as their only child. These efforts were put forth despite having severe rheumatoid arthritis and diabetes mellitus, which caused her pain in many joints and a need to cook special foods separately. She complained of being easily fatigued and was moderately obese. Millie had never lived away from home and had only one brief job as a secretary that she found to be too demanding to maintain given her rheumatoid arthritis symptoms of pain and impaired mobility. Her social life consisted mainly of talking with her girlfriends from her church on the telephone, sometimes 16–20 calls per day to various friends to check on each other. Outings were rare other than going to church-related activities. Millie had suffered bouts of depression over her adult life and was already taking maintenance antidepressant medication. When her mother died, Millie became more depressed and was having a hard time coping with her grief and with maintaining an organized structure to her life, such as the demands of her medical conditions, as well as maintaining a household by herself. It appeared that Millie had already done a good deal of grief work through her pastor and church-related support during the 6 months since her mother died, but it was now dawning on her that these supports were not going to be able to meet all of her own needs.

She was brought to her appointment by her mother's younger sister and her husband (Millie's aunt and uncle), who had visited and helped out regularly during her mother's illness, especially transporting both of them to various medical appointments. In a joint meeting with the aunt and uncle, they painted a different picture of Millie as always being sickly, frail, and needing to be sheltered from many of life's hardships. They also expressed worries that they both had their own medical problems too and were not sure how long they could continue to bring her groceries and take her to appointments. They wondered out loud about her future. Mille had never earned any significant, sustained income; she did not drive and had never taken public transportation. She had real medical problems (poorly controlled diabetes and its complications as well as painful rheumatoid arthritis) and saw herself as a near invalid and was terrified of being left alone to fend for herself.

Millie clearly met the criteria of having dependent personality disorder and when the object of her dependency (her mother) died, she was thrown into a crisis. In getting to know Millie, it became clear that she was reasonably intelligent and that she was completely ambulatory despite her pain complaints. A steep learning curve was ahead of her, however, to learn all the skills she would need to be self-reliant or alternatively, she would need to search for a new person to become dependent upon, but no one appeared to be on the horizon to fill that role. Given the warning voiced by her aunt and uncle,

Millie's IPT therapist chose to begin to encourage Mille to take concrete steps to practice greater self-reliance. One such step was the direct suggestion that she come to her appointments by public transportation instead of relying upon her aunt and uncle to drive her to all of her visits. This suggestion caused considerable distress at first and more than a little skepticism from Mille as well as her aunt and uncle. When the "how to" details were reviewed repeatedly and the "launch date" was set, Millie was, in fact, able to follow through and achieve a milestone of success in making her first independent trip. She had achieved the first of a myriad of "first steps" that would be required to achieve successful independence, something she had never been able to accomplish before age 61. Millie continued to make steady but slow progress and due to her history of repeated episodes of major depression, she agreed to enter long-term maintenance treatment on a monthly basis.

Case Vignette 4: Discussion

This vignette could be repeated with a whole range of disabled individuals who live with aging parents until their parents become too impaired to care for them or until their last parent dies. Sometimes, these needs are anticipated and contingency plans are drawn up ahead of time. In other cases, like Millie's, a crisis ensues.

Not every disabled person who loses his or her primary support will meet criteria for a personality disorder or for the IPT problem area of interpersonal deficit. Even though her mother's death had precipitated her worsening depression, grief did not seem to be the issue that was driving Millie's depression at the point she was first seen, rather it was fear of being unable to cope on her own without inordinate input and help from others. The IPT therapist in this case needed to size up the overall situation and make some strategic decisions about where to encourage Millie to begin to learn the skills she needed to become more independent. In cases of a problem area focused on interpersonal deficit, the therapist often takes a more active stance and uses the therapeutic relationship more directly, by strongly encouraging small steps of progress in other areas of independent functioning. A slow process of change is the usual scenario. Seeing herself as having a string of successes had the effect of bolstering Millie's self-esteem and confidence to try even more challenging tasks. Setbacks, demoralization, and manipulation of potential new objects of dependency are also subsequent developments that needed to be contended with. The relationship with her aunt and uncle, her only remaining relatives, appeared to have left them exhausted and they were signaling that they could not take primary responsibility for her any longer. In Millie's case, she was eventually able to continue to live independently with some help from her friends and her church but she learned to arrange for her own transportation and to manage her medical needs and finances without regular input from others. A referral to a social worker concluded that her physical limitations merited an application for disability. Her mood simultaneously improved and stabilized with these interventions.

These preceding case vignettes illustrate the four foci or problem areas of IPT as it is delivered to individuals over age 60 in the traditional sense, that is, when subjects do not have the complications of coexisting cognitive impairment and who are able to fully participate on their own without caregiver involvement. The interested reader can refer to the writings of the original developers of IPT for more details on techniques and additional case vignettes, as well as the excellent book-length overview of IPT applied to cognitively intact elders (Hinrichsen, 2007; Hinrichsen & Clougherty, 2006). A video about using IPT for geriatric patients is also commercially available for viewing (Hinrichsen, 2007).

Preparation for IPT-ci

Before turning our attention to the modifications to IPT used for accommodating the needs of the patient who shows cognitive impairment, we will first digress in the next five chapters beginning with a review of the evolution of traditional IPT to IPT-ci (IPT-cognitive impairment). The other four chapters deal with gerontology/geriatric medicine, late-life depression, cognitive impairment (particularly executive dysfunction), and dementia. These chapters are intended to be concise summaries of salient facts that will present themselves repeatedly when working with this population and whose combination can cause a variety of challenges and complications. In order to be able to provide ample psychoeducation for identified patients and caregivers, the IPT-ci therapist needs to understand these basic concepts. Graduate-level courses on gerontology are highly recommended. After these topics are presented, Section II will focus on their integration into the specific techniques for modifying IPT for patients with cognitive impairment and their caregivers.

2

Overview and Background for IPT-ci

The principles of IPT as outlined in Chapter 1 will serve as a background for understanding how IPT-ci developed from IPT. This chapter summarizes the adaptations that make up IPT-ci. It will be followed by concise chapters on general principles of gerontology/geriatric medicine, late-life depression, cognitive impairment/dementia, and executive dysfunction (ED) in order to be able to understand how all of these link together and to prepare the student of IPT-ci with a minimum database from which to begin the psychoeducation of identified patients and caregivers when implementing IPT-ci as outlined in Chapters 7–12. The interested reader is referred to the list of references for the original published manuals of IPT and several other resources.

As cognitive impairment is often difficult to assess without a battery of neuropsychological testing, family members (and eventual caregivers) are often not aware of the more subtle aspects of cognitive impairment, particularly executive impairment. In fact, one study carried out by the author's colleague Mary Ganguli, MD, in her study of dementia rates in a population within communities in the steel producing region along the Monongahela River near Pittsburgh showed that 80% of households that contained an elder who met all criteria for a dementing disorder were not recognized as having any cognitive disorder by family members who lived with them (Ganguli et al., 2006). More often than not, cognitive impairment, particularly ED, is misinterpreted, misunderstood, or less commonly, reacted to punitively by family members who have no experience or education to know what they are observing. It is for this reason that this manual will spend considerable time outlining the clinical approach to working with individuals with cognitive impairment, particularly in the early stages (with or without depression).

How Can Psychotherapy Build Upon Itself When There Is Memory Impairment?

Sigmund Freud was initially pessimistic that older patients could benefit from psychotherapy as they were too set in their ways to be able to change in the later phase of life. Fortunately, his successors proved him wrong. In fact, many writings on the issue describe older patients as eager to enter a therapeutic relationship to help them to achieve the meaning they seek as they become more aware that time may be running out for them. Older patients often have fewer inhibitions to reveal personal information compared to younger patients and often assume the attitude "what have I got to lose" (by opening up and being totally honest).

Patients meeting criteria for dementia with severe short-term memory loss would not be considered, by most, to be good candidates for psychotherapy as they cannot build upon prior gains due to their impaired memory. The capacity for insight is also linked to intellectual capacity as a higher order cognitive function that requires an ability to weigh current perceptions and sift through large amounts of stored memory and then apply tests of logic to the comparison based on one's accumulated experience. This ability for insight and reasoning, as well as sophisticated linguistic communication skills are the three main capacities that separate *Homo sapiens* from other species in the animal kingdom. Thus, if the capacity for insight is diminished by diseases that disrupt one or more brain functions, such as strokes, or neurodegenerative processes like Alzheimer's disease, one could reasonably conclude that the ability to benefit from psychotherapy would also be diminished.

On the other hand, Naomi Feil has written a great deal about her work with severely cognitively impaired subjects using Validation Therapy (Feil, 2002), where she puts her psychotherapeutic skills to work to try to infer what is going on internally with severely demented patients who may be emotionally distraught or acting out but unable to explain it to another person. Her background database about the patient's life is used to gather clues that help her to formulate hypotheses to explain the witnessed behavior and formulate a plan of action to try to remedy the underlying need that precipitated the behavior. For example, if a woman with Alzheimer's disease in a nursing unit is exhibiting constant disruptive vocalizations, calling out for something with an unintelligible name, Dr. Feil might gather more history and take note that when the woman was a young adult, she lost a child. On a very basic level, Dr. Feil might hypothesize that this tragic loss has been an overriding preoccupation during this woman's life and now that she has lost many of her compensatory abilities due to her dementia, she is overwhelmed by the pain of that loss. Dr. Feil's intervention might be simply to hug the woman to try to comfort and calm her in a basic tactile way or to give her a furry stuffed animal to hold, which can symbolically represent her lost child in a primitive way. Later on, it might be noted that the distraught Alzheimer's disease victim calmed down and stopped yelling for the unnamed person and is noted to be clutching her stuffed animal tightly and seems to be comforted by it. This hypothetical vignette is presented here to illustrate that the willingness to intervene psychotherapeutically can be beneficial, even with a severely demented person, in some cases.

Traditional psychotherapy requires that rapport be established for an exchange of ideas, thoughts, and feelings. It requires a motivation for change, usually to escape some discomfort and it requires that the therapist be viewed with some hope that their expertise will be beneficial to justify the required effort (and cost) of participation. If we accept that patients with severely impaired memory and insight are not good candidates for traditional psychotherapy but may benefit from some attempts to reach them utilizing psychotherapeutic understanding (as in Validation Therapy), then there is a broad spectrum of eligible candidates who lie between cognitively normal individuals and severely demented individuals who could benefit in some way from psychotherapeutic intervention. The availability of interventions such as traditional IPT can be broadened by utilizing techniques that affect the patient not only directly but also work indirectly through improved understanding and coordinated changes made by the people in the patient's sphere who want to be helpful, that is, the caregivers (as in IPT-ci).

Patients who show both cognitive impairment and depression initially find it difficult to engage in the therapeutic process, establish a focus, and recall the content of former sessions, compared to depressed elders without cognitive impairment. This is not the end of the story, however. In our research, cognitively impaired patients were able, over time, to form a connection with their IPT therapist such that they came to view them as advocates and often looked forward to the sessions. When a context was supplied by the IPT-ci therapist to remind them of work from prior meetings, they were, as a group, surprisingly able to re-engage on the problem being worked on. Some patients expressed hope that the IPT-ci therapist would be a useful go-between to soften the power or reach of what they perceived to be an overbearing spouse or adult child. Distinct from work with cognitively intact patients, the IPT-ci therapist for these patients quickly learned to "launch into reminders" at the beginning of each session in order to use the available time more efficiently. Sometimes, using written reminders or notebooks prompted the identified patients and their caregivers to revisit issues between IPT-ci sessions and carry out implementation strategies more regularly.

In the Maintenance Therapies in Late-Life Depression Study-II (MTLD-2; Reynolds et al., 2006) discussed in the Introduction, subjects over age 60 with major depression, some of whom also had mild to moderate cognitive impairment (mini–mental status scores as low as 18/30), were treated in the acute phase with a combination of both IPT and antidepressant medication (paroxetine). There was no statistical difference between the cognitively normal and the cognitively intact subgroups regarding the proportion who remitted from their depression with this combined treatment. Since all subjects received combination treatments, the singular effect of the IPT cannot, however, be distinguished. One interpretation of the fact that combination treatment worked equally well for both groups (cognitively normal and those with mild to moderate cognitive impairment) might be that the effect of the antidepressant drug was so powerful compared to the IPT such that all the improvement was due to a drug effect in both groups. Alternatively, if one assumes there was some contribution of the IPT to depressive symptom resolution, then it must be concluded that IPT/IPT-ci worked equally well for these two groups.

IPT for Cognitive Impairment

Although this adaptation of IPT is intended to be a useful toolset for working with depressed elders with cognitive impairment, the same principles will apply to engaging and understanding patients with a syndromal dementia (with or without depression), particularly when the issues brought by the caregivers who accompany them suggest evidence of ED. Many of the basics of engaging and working with cognitively impaired elders described here will overlap with the extensive literature that focuses mostly on the management of individuals who already have dementia (Gallagher-Thompson & Thompson, 2006; Gallagher-Thompson, Steffen, Thompson, 2008; Gwyther, 2004) (Burgio et al., 2001; George & Gwyther, 1986; Gitlin et al., 2003; Hinrichsen & Pollack, 1997; Schulz & Patterson, 2004; Teri, McKenzie, & LaFazia, 2005). The particular strength of IPT-ci lies in its flexibility to encompass all individuals on the continuum from normal (no cognitive problem) to syndromal dementia, particularly in the way in which it is designed to educate the caregivers about both cognitive impairment/dementia and depression and therefore equip them with the understanding and new strategies for coping better with the issues at hand as well as preparing for possible changes in the future, particularly further cognitive decline over time.

Among the population of elders who suffer from cognitive impairment, the engagement and education process used by clinicians who are interested in helping those elders and their families to cope maximally is essentially the same whether or not depression is part of the presentation. Patients and family members need to be engaged, evaluated, and educated about what the therapist has come to understand about the nature of the problems they present with, whether the presenting problem is "pure" depression or cognitive impairment or a clinical presentation with elements of both. In evaluating a new patient brought for evaluation by an adult child, for example, it requires the same effort to engage both the identified patient and the caregiver(s), elicit a complete history without alienating the patient, and oftentimes to piece together a composite picture over time of whether the problems encountered are due to depression, some level of cognitive impairment, a combination of both, or to an undiagnosed medical problem. When considering a person who is clearly experiencing cognitive impairment, the presence of depression certainly complicates the picture. If depression is not evident at the outset of the engagement with a cognitively impaired patient, there is certainly a high risk for developing it (50% or greater; Jeste et al., 1988; Reding, Haycox, & Blass, 1985; Zubenko, Moossy, & Kopp,1990) moving forward if the observed cognitive impairment is due to the early stages of a degenerative dementing process such as Alzheimer's disease or Parkinson's disease.

The modified form of IPT dealt with in this manual, hereafter referred to as IPT-ci, is designed to provide ample psychoeducation (in this case about depression and/or cognitive impairment/dementia), an evidence-based treatment for depressive symptom relief, a forum for dyadic role-dispute resolution, a mechanism to engage caregivers who are going through their own role transition, and the consideration of alternative problem-solving solutions for elders who are suffering from cognitive impairment with or without depression. The most frequent problem area noted in depressed, cognitively intact geriatric patients, from our research experience, was some form of "role transition" such as adjusting to retirement. Cognitive impairment can also be looked upon as "role transition" in that it often leads to declining social interaction, declining participation in pleasurable activities, and the development of a self-view as someone with a failing brain (or worse yet when the cognitive impairment is misperceived as a character flaw). These role transitions can lead to demoralization and depression. The techniques of IPT-ci can help victims of cognitive impairment to improve coping skills and perhaps to avert concomitant depression.

Engagement of Caregivers

Beyond psychoeducation for caregivers, IPT-ci offers a forum for empathic understanding of the problems they face in caregiving, a practical forum for role dispute resolution, and an ongoing resource for coordinating the care for the identified patient over the long term, where the development of depression is a high risk and when further cognitive decline in the identified patient is, unfortunately, the likely scenario. The goals for engagement are to make an accurate diagnosis by searching for possible causes of the cognitive decline, to educate the afflicted individual as much as he or she is able to comprehend, to educate the family members or caregivers about the true nature of the disabilities of the individual with cognitive impairment, and finally to assist caregivers to adopt realistic expectations as well as to diplomatically prepare them for what could be a downward trajectory of functioning. The empathic recognition that caregivers may also be going through their own role transition in their attempts to adjust to the implications of the behavioral changes they are noticing is a key modification of IPT over its traditional use. Helping the caregiver to implement strategies for coping optimally in light of the newly comprehended realities is a further extension of IPT for cognitive impairment (IPT-ci). Successfully helping caregivers to find and implement new strategies for coping in realistic ways will indirectly benefit the identified patients with cognitive impairment as they are held to more realistic expectations and treated with more empathy and dignity, despite their limitations.

A vast literature has documented the burden of caring for depressed (Schulz & Patterson, 2004) or dementing relatives (2006) with high reported rates of depression, substance abuse, and worsening medical burden in caregivers. Successfully implementing IPT-ci will hopefully modify these outcomes for the betterment of both the cognitively impaired individual (with or without comorbid depression) and their caregivers or concerned family members.

Although all accompanying concerned family members may not be currently carrying out traditional caregiver roles such as helping with instrumental activities of daily living (IADLs) *per se* in the earlier stages of the identified patient's cognitive impairment, an argument can be made that even increased surveillance for possible problems in the identified patient indicates that they are beginning to take on the role of a caregiver. They are beginning a role transition themselves from purely give-and-take, mutually reciprocal interactions, to more unilateral thinking about what might be going on with the identified patient, to increased surveillance for problem behaviors, to outright assistance (with transportation or finances, for example). Caregivers experience the necessity of this shift with a range of emotions from altruism, sadness, or matter-of-factness, to feeling abandoned, betrayed, resentful or enraged. Throughout the rest of this manual, the term "caregiver" will thus be used for brevity to represent the entire spectrum of individuals who harbor concerns for the mental state or functional capacity of the identified patient with cognitive impairment.

When an older patient is first evaluated in a mental health setting, it is often like a mystery unfolding with a cast of characters, a storyline, sometimes deliberate omissions or concealments, and a complete picture that slowly emerges over time to reveal the essence of the struggle that brought the patient for evaluation in the first place. If the mystery unfolds to reveal a cognitively intact patient who is depressed and a good candidate for traditional IPT, then other parties need not be involved from that point forward for a successful outcome (resolution of the depression). However, the IPT-ci "toolset" is ready to fully recognize the problems being faced by the caregiver(s) on all levels (e.g., practical, personal, historical, and potential role disputes) and it is intended to help the caregiver to be an optimally functioning caregiver for the ultimate benefit of the individual with the cognitive impairment or depression, while respecting the caregiver's right to privacy.

The IPT-ci therapist may recognize the need in caregivers for their own tailored supports that might include their own psychotherapy, and he or she is then prepared to refer caregivers elsewhere to find appropriate help. In other words, the IPT-ci format is a mechanism to work with and help caregivers to cope better with their caregiving role without making them psychotherapeutic patients in their own right (with the same therapist). If it is determined that caregivers could benefit from their own psychotherapy, a referral to a colleague within the same setting that allows for simultaneous treatment for both caregiver and patient during their visits would be optimal. This strategy helps to maintain a thread of clarity in the mind of the treating IPT-ci clinician, who may feel overwhelmed at times with the multiplicity of needs expressed by both the identified patient and their accompanying family members/caregivers. Like a lighthouse in a storm, the IPT-ci therapist is reminded to maintain their primary goal of serving as an advocate for the identified patient's best interests. They may also be simultaneously helpful and useful to the accompanying caregivers as a result of this effort but, should the waters become muddied with the needs or demands of the caregiver, the allegiance of the IPT-ci therapist should seek to steadfastly advocate for the best interests of the identified patient. In the rare instance, this might require taking action against the caregiver if there is suspicion of elder abuse. In practice, balancing the needs of the identified patient through individual advocacy as well as listening to caregivers and helping them to problem solve will be the bread and butter of the IPT-ci therapist, whose goal is to seek the highest quality of life and functioning ability for the identified patient by minimizing his or her depressive symptoms, helping him or her to cope maximally, and to accept with dignity the cognitive impairment that cannot be further remedied after an adequate medical work-up has been carried out and appropriate trials of cognitive enhancing therapies have been maximized.

Delivering IPT-ci in Common Care Settings

Regarding IPT-ci, where the focus is on the identified patient with cognitive impairment with or without depression, some digression about usual care is in order. As mentioned earlier, cognitive impairment often goes unnoticed even when it is at the level of a diagnosable dementia and even more so in its earlier stages such as mild cognitive impairment (MCI). Memory loss as evidenced by forgetting names, appointments, and recent events or continually losing objects or one's way in familiar places are the usual harbingers of cognitive decline noticed by family members whether or not the identified patient notices them too. Executive dysfunction (covered in greater detail in Chapter 6) is actually more prevalent than memory decline but is less recognized such that it has been called a silent epidemic by some experts in the field. Evidence for ED [not to be confused with the other ED—erectile dysfunction] can be variously noticed as socially inappropriate behavior (uncharacteristic swearing, insensitive remarks), quick temperedness or impulsivity (e.g., road rage, demandingness, or increased self-centeredness), poor problem-solving skills (e.g., difficulty using appliances, making repairs, and driving safety errors), impaired insight (inability to see multiple or complex components in a situation), and poor multitasking (poor ability to shift attention such as when continuing to prepare a meal or drive a vehicle while talking on the phone simultaneously). Although every normal adult has lapses in one or more of these areas from time to time, a repeating pattern of examples from multiple categories can suggest ED, as the case vignettes in later chapters will illustrate. Normal adults take these skills for granted and as deficits in these areas usually come on gradually, they often go unnoticed until the examples become extreme or a crisis ensues (such as a call that the afflicted cannot find his or her way home in familiar territory or after experiencing an embarrassing social faux pax).

Complaints about suspected cognitive impairment are most commonly initially brought to primary care physicians, who may suggest a medical workup, a trial of a cholinergic enhancer, or a referral for neuropsychological testing, or to another specialist like a neurologist or psychiatrist, not necessarily in any order. If depression is also evident, a referral to a psychologist, psychiatrist, or social worker might also be made but are commonly not followed through with by the patient. The reasons for not following through with referrals are multiple and include the perception of stigma in seeking mental health help, transportation complications, local availability, and cost. With a typical brief encounter in a primary care setting, the patient and family may not have had enough time to truly understand all the implications of what was being conveyed (inadequate psychoeducation). Treatment for depression in most primary care settings is commonly limited to trials of antidepressant medication.

Comprehensive geriatric centers often have access to all the aforementioned services under one roof but are not accessible to all patients. For those individuals who show progressive cognitive decline, primary care physicians (or the nurses, nurse practitioners, or physician assistants working with them) are sometimes the only healthcare advocate working with these patients until late in the course when daycare, respite care, in-home services, or nursing home placement options are deemed necessary to explore. In geriatric specialty care, a social worker often serves the role of working with family members to try to educate them and to provide them with resources within the community to explore services such as day care, senior centers, and so on as they are needed. Social workers sometimes provide supportive psychotherapy or specialized dementia care services as well. Psychologists sometimes consult directly with primary care practices and some even have regular hours to see patients on-site. Most psychologists, however, are in private practice (if there is local availability at all) and thus a willingness to follow through in a different setting than the primary care doctor's office is required.

It is also important to note that the current cohort of older individuals may have little or no experience with the concept of psychotherapy (or be willing to pay for it) as some have no prior experience or prior exposure to psychology classes in high school or college. Many current elders have also had a lifelong history of coping without outside help and do not see themselves as the type of person who looks outside their immediate experience to solve "personal problems," with the possible exception of confiding in their pastor, priest, or rabbi.

Who Should Implement IPT-ci?

The use of collaborative models of healthcare has clearly improved outcomes for patients with asthma, heart disease, and diabetes beyond what a single practitioner could achieve. Similarly, in the PROSPECT (Bruce et al., 2004) and IMPACT studies (Unutzer et al., 2002), depression care specialists working within primary care settings achieved more complete resolution of depression. The principles of IPT-ci can be taught to these same depression care specialists to equip them to be able to fully engage older patients with cognitive impairment/dementia within primary care settings and subsequently to work with the patient and caregivers to maximize treatment in the short and long term.

In the IPT-ci model of care, service would be provided by a nurse, social worker, psychologist, psychiatrist, or primary care physician (although the latter two are more likely to have supervisory roles over the patient's overall care and have the IPT-ci carried out by one of the first three). If the IPT-ci therapist is functioning in an independent capacity, medical and/or psychiatric evaluation and management on a consultative or preferably a collaborative basis would be required to insure that an adequate medical evaluation was carried out to look for treatable causes of cognitive impairment and depression as well as to evaluate and manage the appropriate use of psychotropic medications when indicated. The use of psychotropic medications for depression, cognitive enhancement, anxiety, agitation, insomnia, oppositional behavior, or psychosis is sometimes indicated at various stages of cognitive decline and is completely compatible with the concomitant use of IPT-ci. Referral to a psychologist or neuropsychologist or sometimes a neurologist may be indicated to provide neuropsychological testing batteries or other specialized diagnostic testing to make a definitive diagnoses of the cognitive impairment as best as it can be determined. Usually the best that can be determined is a "probable" diagnosis given that a brain tissue biopsy may be the only way to make a definite causal attribution. We will explore dementia work-ups more in Chapter 4.

If social workers were delivering the IPT-ci, it would be entirely appropriate to integrate their other social work duties into the delivery of IPT-ci, such as providing lists of community resources, pursuing financial assistance, and assisting with arrangements to procure power of attorney documents and living wills and so on. In fact, social workers, particularly those who already work in multidisciplinary geriatric care facilities, are ideal candidates to utilize IPT-ci to the fullest. If a master's level nurse or clinical psychologist is delivering the IPT-ci, they may need consultation with a social worker to explore community resources or other services they may be less familiar with. Any of these professionals could provide the patient and caregiver with the advocacy outlined in this IPT-ci manual. With the demographic trends of an aging population with greater numbers of expected victims of cognitive impairment, the creation of a specialty training program specifically for geriatric care managers using IPT-ci principles as a core structure for providing clinical care may be called for in the near future.

3

A Concise Review of Gerontology/Geriatric Medicine

For the clinician who is beginning his or her work in the geriatric population, an adequate background in gerontology/geriatric medicine would obviously be useful to put what is being observed into proper perspective based on the work and cumulative experience of predecessors in this field. A thorough course in gerontology at the master's degree level is highly recommended. This chapter will outline a few basics that the author has found to be of universal usefulness to review as a pretext to implementing IPT-ci.

Background Information on Gerontology

The study of gerontology or aging is a relatively new field. With rare exceptions, the average lifespan of humans throughout history has peaked from the fourth to sixth decade of life. It hasn't been until the twentieth century that the average lifespan in the industrialized world has expanded to nearly eight decades. Currently, the longest living human on record was 122 years.

To prepare the reader for common maladies associated with aging, we will concentrate on age-related degenerative diseases. As public health practices such as better sanitation, water purification systems, immunization programs, and advances in the practice of medicine took place, the average lifespan has continued to edge higher. As a population, we are not dying as often from heart disease, cancer, strokes, and infectious diseases, but we are living long enough to accumulate non–life-threatening chronic problems such as painful arthritis, hearing and visual problems, and Alzheimer's disease. These chronic problems can produce disability and impaired function in IADLs such as cooking, shopping, cleaning, and managing personal finances effectively. With increasing disability, more basic ADLs (such as dressing, bathing, toileting, and transferring) can be affected. These disabilities can be thought of as a continuum of severity that may require some assistance from others to remain living independently as long as possible. The needs for care have spawned a whole array of services such as meals on wheels, visiting nurses, subsidized taxi services, and home health aids (to assist with bathing, for example). Housing options have changed from living out one's life in the proximity of adult children to such options as assisted living facilities, life care communities, and the subsidized senior high rises that provide dignified living arrangements for elders, particularly when adult children have moved away to find work in a global economy. Adult children sometimes try to monitor and sometimes micromanage the aging parents' needs by telephone or through the help of local agencies that specialize in these types of care when they live in other locations.

Comprehensive geriatric health services have sprung up to meet a variety of health care needs for seniors. These services have learned the value of coordinated care across disciplines for the whole person, particularly caring for the frail elderly who have historically fallen through the cracks when no single health care professional was keeping an eye on the big picture.

The Aging Process

To digress for a moment on the nature of the aging process, several theories have been put forth, such as the accumulation of errors in our DNA from ionizing radiation, or the cross-linking of collagen which leads to stiff joints and thin skin as age progresses. When considering that other species in the animal kingdom have varying life spans that are species specific, from a few days for fruit flies to 200 years or more for some tortoise species, the aging theory that makes the most sense to the author is the one called programmed cell death. There is even a formula for the rough calculation of this phenomenon, which states that the average lifespan across species is roughly nine times the period of time from birth to sexual maturity. If we assume this theory to be true, and presuming an average age of puberty of 13 years, 117 years would be the maximum lifespan for *Homo sapiens* without life-shortening medical- or stress-related events.

Physiologic Changes of Aging

Even though we, as a population, are now living longer, disability and chronic medical burden can profoundly affect our quality of life and increase the risk for depression. There are several physiologic changes of aging that are important to note. The acuity of all five senses regularly diminishes with age. Hearing loss is common and, unfortunately, hearing aids are too often perceived as annoying and are thus underused because they amplify background noise as well as speech. The most advanced hearing aids use sophisticated computer chip circuitry to filter background noise, but these are very expensive to obtain for average income elders. Sight can also decline associated with aging for a variety of reasons such as cataract formation (lens opacification), macular degeneration (degenerating retinal function), glaucoma (elevated intraocular pressure), or diabetic retinopathy (vascular abnormalities in the retina). Some drugs, like Viagra®, for example, can affect vision by giving it a blue tint. The sense of smell and taste are linked and can degenerate from damage to the nose (secondary to a fracture, for example) or from

Alzheimer's disease when it causes degenerative effects in the olfactory cortex of the frontal lobes. An impaired sense of smell is one test neurologists use to evaluate a possible dementia diagnosis by asking the patient to identify the smell in small vials that contain coffee, cinnamon, or lemon extracts. Impaired taste can also result from some drugs; for example, Lunesta® can cause a fishy taste in about 20% of people. Impaired smell and taste senses can put an elder at risk for missing a gas leak or eating tainted food. Alternatively, impaired taste and smell can lead to a decreased appetite, which can result in weight loss and weakness. Skin loses collagen with age and becomes thinner, sometimes like tissue paper. An elder's skin can tear easily with minor trauma and as pain sensitivity can also decline with age, an elder could be at increased risk for burns from scalding bathwater or from lying too long on a heating pad.

With aging, there is a decline in total body water such that drug doses often need to be adjusted downward to avoid toxicity as a person ages. Furthermore, every organ system can lose efficiency with aging. For example, the liver may become less able to metabolize drugs because the aging heart pumps less strongly and cannot push the blood through the liver to carry the drugs to be metabolized as efficiently. The aging kidneys also become less efficient in getting rid of drug metabolites and toxins. The net effect of these developments is a greater sensitivity to drug toxicity. When you couple this fact with the frequent occurrence of polypharmacy among elders, it is no surprise that elders can become confused, dizzy, and at risk for falls due to drug toxicity.

Polypharmacy

Polypharmacy can begin with the good intentions of a doctor who is focusing on one problem without taking into account the cumulative effects of all medications the patient is taking. Sometimes legitimately prescribed medications can also work at cross purposes and can cause harmful drug interactions such as one drug inhibiting the normal degradation of another drug resulting in a very high blood level. Of course, over-the-counter medications, shared drugs, and recreational drug abuse can all increase the risk for toxicity. A good practice is to constantly seek to simplify drug regimens to avoid this problem.

CASE VIGNETTE 5: ILLUSTRATING A DRUG–DRUG INTERACTION

As a geriatric psychiatrist, the author was referred a patient from a neurologist. The patient had an inoperable brain tumor and was sent for treatment of his accompanying depression. When meeting this gentleman, he reported worsened motor coordination in the past 2 days such that he had to hold onto walls to walk. It turns out that this new symptom was not due to his brain tumor but a drug interaction. His neurologist had prescribed the antiseizure medication phenytoin (Dilantin®), which is commonly prescribed as brain tumors can predispose to seizures. He also added a particular antidepressant medication 3 days earlier, which as it turns out, decreases the metabolism of phenytoin. The net result was that when the blood level of phenytoin was checked, the previously therapeutic dosage was now in the toxic range. A prominent side effect of phenytoin toxicity is gait instability.

Effects of Medication and Other Drugs in the Elderly

When elders are admitted to our inpatient psychiatric unit in a state of confusion, it is frequently the case that when all nonessential medications are stopped, they often clear mentally in a day or two. To compound this problem, over-the-counter medications such as antihistamines or common cold remedies can contain ingredients that can lead to mental confusion or delirium, a topic we will explore further in Chapter 4. Alcohol, nicotine from tobacco use and caffeine, or their sudden withdrawal, can also contribute to psychiatric disorders such as depression, delirium, or chronic insomnia. Finally, poor compliance with medication regimens due to forgetting, self-manipulation of doses for perceived side effects, or for cost savings can all contribute to underdosing and forgetting and then retaking the same dose or concluding that more will be better (e.g., for controlling pain) can lead to overdosing. Patients have also been known to share their medication or to "try out" the medication of their neighbors because it helped them and they wonder if it will do the same for them. Given the extraordinarily high cost of medications, some elders have resorted to buying medication from overseas suppliers whose drugs may not meet FDA standards for safety and efficacy and in some cases, such pills have been found to have no active ingredients (which could be a fatal unwitting omission for a congestive heart failure patient, for example).

Alcohol and drug abuse can occur at any age and the broad group of elders is no exception. Retirement and increased leisure time is often conducive to more frequent use of alcohol that can result in a net increase in consumption for a given elder. Long-standing drinking habits such as "cocktail hour" before dinner can become a toxic ritual as the individual ages or is in the early stages of Alzheimer's disease. These individuals often don't realize that the same drink volume they have been consuming for 20 years now causes slurred speech or a wobbly gait. Alcohol rehab services may be required for elders as much as for younger individuals. A few drug and alcohol rehab facilities have specialized services just for elders.

Sleep Problems in the Elderly

Insomnia is a frequent complaint among the elderly (Driscoll et al., 2008). Research has shown that the stages of sleep do become altered with the aging process and sleep becomes more fragmented and less consolidated. Less total sleep time, less deep sleep (the most restorative kind), and more frequent awakenings are also the rule more than the exception. Many elders also must rise from sleep to empty their bladder. Men can have prostate gland enlargement, which put pressure on the bladder giving a sense of urgency to urinate. Women, particularly those who have delivered children from vaginal births, often suffer from incontinence due to stretched pelvic tissue. Urinary incontinence also puts both genders at risk for urinary tract infections. Both genders can also show nocturia (frequent nighttime urinating) due to the use of diuretic medications (water pills), particularly if they consume their last dose late in the day. These sleep complaints, unfortunately, often result in the prescription of a sleeping pill that adds another psychotropic agent.

The newer sleeping pills are less habit forming and have less "hangover" effects than the purely sedative drugs used in the past but all of them can induce psychological dependence, increased risks for falls (especially at night), increased risk for sleepwalking (somnambulism), and memory loss or confusion in vulnerable individuals. Other common causes of poor sleeping are emphysema, heart disease, alcohol abuse, restless leg syndrome (involuntary leg jerks that can wake a person into a lighter stage of sleep or to full wakefulness), and sleep apnea. Sleep apnea is defined as pauses in breathing during sleep that lead to lowered oxygen levels in the blood, which then triggers an internal alarm causing the person to gasp for breath and awaken to a lighter stage of sleep. These sleep apnea cycles can repeat all night long, robbing the person of restorative sleep and leading to excess daytime drowsiness. The treatment for sleep apnea involves keeping the airway open with a device that keeps the pressure in the airway high enough to ensure regular breathing (CPAP or continuous positive airway pressure).

Other Issues Related to Older Age

From a medical standpoint, older age can also herald the onset of a multitude of medical problems that deserve acute care and regular follow-up to ensure that the treatments for various ailments are balanced and take the whole person into account. The risk for falls (resulting in hip fractures or worse), medication errors, adverse effects of alcohol or other drugs, and vigilance for the onset of depression and cognitive impairment need to be maintained throughout with appropriate regular follow-up intervals to gauge progress or to uncover possible new problems.

Erik Erikson, a famous developmental psychologist, described the last stage of life as an internal struggle between ego integrity versus despair

(Erikson, 1950) to indicate a period in life where some are able to draw upon their life experience in an integrative way more than others. Some elders are able to take stock of their accomplishments and failures, to feel relatively secure about their place within the universe and the human experience, and to be able to reflect on their journey through life and put words to their experience such that others might benefit as well. The opposite pole of despair would characterize those full of doubt, regret, or bitterness when they look back over their life. As advocates or IPT-ci therapists for older adults, of course the former scenario is the preferred one to encourage.

Importance of Social Support for the Elderly

From a social support standpoint, the adequacy of a social network has a protective buffering effect against depression and the lack of one can leave an elder feeling vulnerable or even helpless as they age into a frail condition or have medical or cognitive problems that interfere with their ADLs and IADLs. Concerned family members can be extremely helpful but may not be readily available due to other obligations (job, childrearing, etc.) or due to geographic distances. Other social agencies have sprung up to fill these gaps such as senior centers, meals on wheels, respite care, in-home companions, visiting nurses, and an array of specialized medical services such as intravenous home therapy and in-home physical therapy.

For individuals who are showing signs of cognitive impairment, particularly those with poor insight that they have a problem, these scenarios may require family members, such as an adult child, to choose safety over their parent's autonomy by insisting on making the decision on a move to live with an adult child or into an assisted living facility, personal care home, or other long-term care facility. In the worst-case scenario, guardianship may need to be pursued to protect a vulnerable older person who can no longer manage their basic needs. We will return to this and other legal issues in Section II.

Contributions of Older Individuals

So far in this chapter, it seems as if we have chronicled a series of losses that require treatment or extra help. Let us end this chapter on some positive note. Even in a culture such as that in America, where youthfulness is celebrated, the unique contribution of older members of society is becoming more and more evident. The over-sixty crowd is more educated than ever before and their years of collective wisdom go beyond mere years of formal education. Many retirement communities are essentially "brain trusts" comprised of retired community leaders, retired professionals, and business people. They have wisely chosen to plan for their golden years all the way to the point of death such that they will not have to depend on their children for help or decision making as they age. Instead, they utilize their accumulated financial assets for preplanned living arrangements that will meet their changing needs as they age, including end-of–life care in a nursing home environment if necessary. This may be the ultimate expression of independence with thought-through contingencies that include such provisions as power of attorney arrangements, long-term care insurance, and advanced directives that help to insure these outcomes. Many of these individuals take advantage of condominium-like amenities and remain engaged in a variety of active endeavors such as volunteerism, community service, the pursuit of hobbies, travel, and even consulting work. Many retired business people have also chosen to lend their years of experience to younger entrepreneurs to offer advice and mentoring.

Moving to a life care community requires considerable financial resources, however, and thus will not be an option for those who cannot afford it. Choices are more limited and waiting lists may be longer for the less wealthy, but government subsidy of such things as senior high rise apartments that cap a senior's cost at 25% of his or her income allow for dignified and safe living arrangements for a much broader group of elders.

Due to the compounding effects of years of saving money, a large proportion of the world's wealth rests in the hands of older individuals. This collective financial power allows for discretionary spending as well as legacy giving to children, grandchildren, and charities.

Lastly, it is important to note that with the average lifespan of 80 years or more, it means that someone age 80 in 2008 was born in 1927. These individuals lived through the Great Depression, two world wars, the invention of television and the personal computer, and landing the first human on the surface of the moon. Their age alone makes them a repository of personalized history whose memories can be tapped by later generations to add perspective and clarity to what can be read in the history books.

Summary

This chapter has been deliberately skewed toward the medical needs of older individuals. We have reviewed the sensory and physiologic changes that accompany the aging process, particularly those that effect drug metabolism. We have reviewed theories of the aging process and sociodemographic trends that affect elders in today's American society. Finally, we have tried to balance the big picture by pointing out the resilience and independence of many older individuals and some of the ways in which they continue to contribute to society at the family level, the society level, and at the economic level. For concise book-length reviews of geriatric psychiatry, see Sakauye (2008) and Spar and LaRue (2006a).

4

Overview of Late-Life Depression

"Depression" is a word that has many meanings, such as a dent in an object or an economic downturn, as well as lowered mood. Sufferers of low mood are usually not familiar with the latest nomenclature used by mental health professionals, nor should they be. The diagnostic task for the evaluator is to make an assessment of symptoms and categorize them in his or her own mind to differentiate levels of clinical severity and choose appropriate treatments, which are typically antidepressant medication, psychotherapy, or a combination of the two, and less frequently electroconvulsive therapy (ECT), or one of the evolving alternative stimulation techniques such as deep brain stimulation (DBS) (Mayberg, 2003), vagal nerve stimulation (VNS) (2008b; 2008c), or repetitive transcranial magnetic stimulation (rTMS) (2008a) (Tarsy, Vitek, Starr, & Okun, 2008).

Using the *DSM-IV-TR* conventions, mental health professionals have agreed that the term "major depression" refers to a collection of symptoms that must include either low mood or lack of pleasure (anhedonia) that is present for most days during the past 2 weeks accompanied by other symptoms or bodily dysfunctions, such as sleep, appetite, or weight change, fatigue, agitation or listlessness, poor concentration, feelings of guilt or thoughts of death, such that the total number of symptoms equal five or more (out of nine). See Table 4.1 for *DSM-IV-TR* criteria for major depression.

The overall intensity of symptoms is rated from mild to severe but all levels must interfere with social or occupational functioning in some way to qualify as a major depression (Williamson & Schulz, 1992). A diagnosis of bipolar disorder requires at least one lifetime episode of mania or hypomania.

Psychosis can also result from severe depression and is usually characterized by extremely negative false beliefs such as believing one has terminal cancer or guilt-ridden thoughts such as the belief that one is deserving of punishment or death. More bizarre delusions such as feeling one is being controlled by aliens from outer space is more typical for a schizophrenia spectrum disorder.

Table 4.1 *DSM-IV-TR* Criteria for Major Depression

Five or more of the following 9 symptoms that have been present for at least 2 weeks, are a change from prior function and contains either #1 or 2.

1. depressed mood
2. diminished pleasure (anhedonia)
3. weight loss or a decrease in appetite
4. sleep impairment (insomnia or hypersomnia)
5. psychomotor agitation or retardation
6. fatigue or loss of energy
7. feeling worthless or excessively guilty
8. poor concentration or indecisiveness
9. recurrent thoughts of death

Source: American Psychiatric Association. (2000). *Diagnostic and statistical manual of mental disorders* (4th ed., Text revision). Washington, DC: Author.

Depression in the Older Patient

Regarding geriatric patients, the reader should keep in mind that today's cohort of elders grew up in the first half of the 1900s and may not have had much exposure to psychology classes. More often than not, older patients who are eventually diagnosed with major depression go to their primary care physician complaining of increased aches and pains and seek relief for these somatic symptoms, even when the symptoms "add up" to a typical presentation for major depression. A diagnosis of depression often must be inferred when taking the "whole picture" into account, which includes input from family members, direct questioning of the patient, and objective evidence such as weight loss, and observations of body language such as sluggish movements, stooped posture, slowed responses to questions, or overly negative content to their speech. Older individuals sometimes do not want to hear that they may be depressed and will sometimes deny or disagree with such a conclusion as they interpret it as a sign of character weakness or because they feel such a diagnosis includes them in the same category of mentally ill individuals that include "crazy people." The doctor or treating clinician therefore must sometimes "build a case" to convince a depressed older person that major depression is an equal opportunity disorder that can affect anyone at any given point of vulnerability. The clinician's task is to convince the depressed person to accept treatment and as the treatment begins to work, to continuously point out the connection between symptom relief and their compliance with the treatment.

Causes of Depression

Genetic vulnerability to depression can occur in families, with the risk being greater when more relatives are afflicted and the closer the kinship. Parents, siblings, or children of an identified person with a depressive disorder have the greatest associated risk although it is never a sure thing as the modes of inheriting risk are not well understood and may involve multiple genes as well as other nongenetic factors yet to be discovered. A given individual within a family that contains several depressed members can still be exempted when the deck of genetic cards are dealt (Smoller & Gardner-Schuster, 2007).

The neurobiology of depression is a complex topic that is being explored from multiple angles, such as biochemical studies (Buysse, Hubbard, Ombao, Houck, & Monk, 2001; Kessler et al., 2005; Maletic, Robinson, Oakes, Lyengar, Ball, & Russell, 2007), neuroimaging (Lawrence et al., 2004; Stein, 2008; Jeste, Lohr, & Goodwin, 1988), genetics (Badner & Gershon, 2002; Smoller & Gardner-Schuster, 2007), and sleep studies (Dew et al., 2003).

Medical Conditions and Depression

Medical conditions can commonly predispose one to depression or exacerbate it, including an over- or underactive thyroid gland, vitamin BI2 deficiency, and strokes (Trepacz, Teague, & Lipowski, 1985), most commonly (Miller & Reynolds, 2002) as illustrated in the following case vignette.

CASE VIGNETTE 6: DEPRESSION CAUSED BY A MEDICAL CONDITION

A 60-year-old man was brought by his wife to see a psychiatrist for treatment of depression and anxiety. His symptoms had worsened gradually over the past 3 months, and he was particularly troubled by insomnia, for which he began drinking several glasses of wine nightly to try to get to sleep. He admitted to some stresses at work but denied that they were extraordinary. He complained of feeling very keyed up, nervous, unable to concentrate, and sometimes paced because he "didn't know what else to do."

In observing him up close, he appeared to be sweaty and clammy. His doctor wondered what role the increased alcohol was playing or whether the stress of work was overwhelming him. The clue to the correct diagnosis in this case was the observation that, when he turned his head to the side, the light happened to illuminate veins in his neck in such a way to notice that his heart rate was quite rapid (the noticeable pulsation reflects the pulsing artery beneath the vein). In taking an accurate pulse reading, his heart rate was 150 beats per minute (as high as in someone doing strenuous physical activity). Blood tests confirmed that he was suffering from hyperthyroidism, and immediate steps were taken to arrange treatment with medicine to block the production of more thyroid hormone, as well as other medication to slow his heart rate down. As this treatment continued for several months, he still

CASE VIGNETTE 6 (Continued)

required medication for depression as well and a great deal of support and understanding to cope with his condition. Eventually, he did return to his old self and was no longer depressed or unable to concentrate mentally.

A long list of other medical maladies can also be associated with depression (see Table 4.2). Moreover, a long list of drugs include depression as a possible side effect for at least some people (see Table 4.3). It is the physician's responsibility to be surveillant for any additional symptoms that might indicate that another medical condition is contributing to depressive symptoms and to attempt to treat such underlying conditions accordingly. Chronic pain is universally depressing to live with and the two conditions frequently co-occur (Karp et al., 2005).

Table 4.2 General Medical Conditions With Depressive Features

Cardiopulmonary	
Cardiomyopathy	Congestive heart failure
Cerebral ischemia/stroke	Postmyocardial infarction
Chronic obstructive pulmonary disease (COPD)	Restrictive lung disease
Endocrine	
Hyperthyroidism	Hypothyroidism
Cushing's disease	Addison's disease
Hyperparathyroidism	Hypoparathyroidism
Hypoglycemia	Hypopituitarism
Pheochromocytoma	Carcinoid
Ovarian failure	Testicular failure
Premenstrual syndromes	Menopausal symptoms
Prolactinoma	
Infectious/Inflammatory	
Influenza	Tertiary syphilis
Mononucleosis	Hepatitis
HIV/AIDS	Tuberculosis
Encephalitis	Lyme disease
Systemic lupus erythematosis	Rheumatoid arthritis
Toxoplasmosis	Viral pneumonia
Neoplastic	
Pancreatic adenocarcinoma	Lung neoplasms
Leukemias	Lymphomas

Table 4.2 (Continued)

Neurological
Multiple sclerosis
Head trauma
Wilson's disease
Epilepsy

Parkinson's disease
Brain tumors
Huntington's disease

Nutritional deficiencies
Folate deficiency
Pyridoxine (B6) deficiency
Thiamine (B1) deficiency

Metabolic deficiencies
B12 deficiency
Riboflavin (B2) deficiency
Iron deficiency

Source: Kupfer, D., Horner, M. S., Brent, D., Lewis, D., Reynolds, C., Thase, M., Travis, M., & Horner, M. X. (Eds.). (2008). *Oxford American Handbook of Psychiatry*. New York: Oxford University Press. Reprinted with permission.

Table 4.3   Pharmacological Causes of Depressive Symptoms

Analgesics and Anti-inflammatory Agents

Fenoprofen	Ibuprofen	Indomethacin
Opiates	Phenacetin	Phenylbutazone
Benzydamine		

Antibacterial and Antifungal Agents

Ampicillin	Clotrimazole	Cycloserine
Ethionamide	Dapsone	Griseofulvin
Metronidazole	Nitrofurantoin	Nalidixic acid
Sulfonamides	Streptomycin	Tetracyclines
Thiocarbanilide		

Anticholinesterases

Cimetidine	Diphenoxylate	Lysergide
Mebeverine	Metoclopramide	Salbutamol

Antineoplastic Agents

C-Asparaginase	Bleomycin	Mithramycin
Trimethoprim	Vincristine	Zidovudine
6-Azauridine		

Cardiac and Antihypertensive Agents

Bethanidine	Clonidine	Digitalis
Guanethidine	Hydralazine	Lidocaine
Methoserpidine	Methyldopa	Oxprenolol
Prazosin	Procainamide	Propranolol
Reserpine	Veratrum	

Antipsychotic Agents

Butyrophenones	Phenothiazines

Table 4.3 (Continued)

Sedatives and Hypnotics
Barbiturates	Benzodiazepines	Chloral hydrate
Chlorazepate	Chlormethiazole	Ethanol

Steroids and Hormones
Corticosteroids	Danazol	Oral contraceptive
Prednisone	Norethisterone	Triamcinolone

Stimulants and Appetite Suppressants
Amphetamine	Cocaine	Diethylpropion
Fenfluramine	Phenmetrazine	Caffeine

Neurological Agents
Amantadine	Baclofen	Bromocriptine
Carbamazepine	Levodopa	Methsuximide
Tetrabenazine		
Phenytoin		

Other Miscellaneous Agents
Acetazolamide	Choline	Cyproheptadine
Disulfiram	Interferon	Methysergide
Meclizine	Pizotifen	

Source: Kupfer, D., Horner, M. S., Brent, D., Lewis, D., Reynolds, C., Thase, M., Travis, M., & Horner, M. X. (Eds.). (2008). *Oxford American Handbook of Psychiatry*. New York: Oxford University Press. Reprinted with permission.

Medication and Depression

Prescribed medications can also be associated with or cause depression, particularly some high-blood pressure medications, steroids, sedatives, and some anticancer and immune modulating agents (see Table 4.3). The surest way to check for a drug-induced depression risk is by asking a pharmacist to check his or her database for each drug in question. It is important to keep in mind that the average elder takes five or more medications and sometimes 30 pills a day. Multiple drugs often have cumulative or interacting toxicities, particularly for elders who are more vulnerable to overdosing by virtue of their slowed metabolism and lowered drug clearance, or due to reduced kidney and liver function as outlined in Chapter 3. As humans age, they also show a decline in muscle mass and total body water in which the drugs are dissolved. This causes an increase in the concentration of the drug in the bloodstream much like a jar of salt water that gets saltier as more water evaporates. Psychotropic drugs are also fat soluble and the proportion of body weight which consists of fat is the only physical parameter that increases as a percentage of body weight as humans age. This increase in body fat creates a depot effect for psychotropic drugs that can contribute to confusion, sedation, agitation, and an unsteady gait when they exceed a vulnerable older person's tolerance limit. The risk of drug-induced toxicity is compounded if cognitive impairment or early dementia is developing, a topic we will address in the next chapter. In other words, it takes less drug toxicity to push a dementing person over the threshold into abject confusion than it otherwise would be with a cognitively intact person. In fact, when elders present to an emergency room or a psychiatric unit with confusion, trimming nonessential medications often restores them to mental clarity (or at least their baseline) within 24–48 hr (Spar & LaRue, 2006b).

Alcohol and Depression

Alcohol use can also lead to depression for several reasons. Alcohol acts directly as a central nervous system depressant. Alcoholics can also wake up on a given morning and become depressed when they realize how much damage their drinking has wrought on their relationships, job, or finances. Chronic alcoholics also accumulate medical problems that can run the gamut from gastritis to internal bleeding, liver failure, seizures, falls, head injuries, and increased rates of accidents and injuries, the consequences of which can cause or worsen depression (Oslin, Katz, Edell, & Ten Have, 2000).

For all the above reasons, the correct diagnosis of depression is often missed by patients, family members, and even well-intentioned primary care doctors. Oftentimes, the true picture is best seen by those who step back and take in the broadest viewpoint by including all available sources of information.

Treatment for Depression

The mainstay of treatment for serious depression has been the use of antidepressant drugs, the most common class being the selective serotonin reuptake inhibitors (SSRI). Although the exact mechanism through which these drugs work is not known for sure, they all boost the availability of the one or more neurotransmitters or chemical messenger concentrations in key areas of the brain thus affecting the resultant electrochemical functioning of specific brain circuits (Warden, Rush, Trivedi, Fava, & Wisniewski, 2007). The SSRI class of drugs has become popular as they work fairly well (35–60% of the time trying a single drug) and have relatively few side effects (commonly nausea, headaches, a change in bowel habits, and sexual dysfunction), which not everyone experiences (Warden et al., 2007). Other classes of antidepressant medications also have a track record of success but must be balanced against their side effect profile and the particular symptoms the depressed person shows. For example, if the depression is characterized by lethargy and oversleeping, a more activating drug may be chosen, whereas if they are already agitated or have insomnia, a more sedating drug might be a better choice for the physician or psychiatrist to prescribe.

Talking therapy or psychotherapy has been shown in controlled studies to work as well as antidepressant medications in all but the most severe cases of nonpsychotic, nonbipolar, depression (Arean & Alexopoulos, 2007; Bharucha, Dew, Miller, Borson, & Reynolds, 2006; Mackin & Arean, 2007; Post et al., 2008; Schulberg et al., 2007). Even though antidepressant drug prices are seen by many as too expensive or even unaffordable, psychotherapy is usually more expensive if dispensed by a private practitioner whose rates are typically $100 per hour or higher. Psychotherapy services provided in a clinic setting by salaried employees are generally more affordable depending upon insurance coverage, which is often, unfortunately, restricted for psychotherapy services. Availability and accessibility to effective psychotherapy services is also limited in many locales, particularly rural settings. The underlying stigma of needing to travel to see a mental health professional rather than remaining in a medical treatment setting is another reason cited by many patients for not complying with mental health referrals. Psychotherapy services made available within a primary care treatment setting are more likely to be accepted as noted in the PROSPECT (Bruce et al., 2004) and IMPACT studies (Unutzer et al., 2002). Of all the varieties of psychotherapy that have been available over the past 30 years, there is a recent trend toward the use of time-limited psychotherapies that target symptom reduction rather than whole-person restructuring or personality change. Evidence-based short-term psychotherapies have been shown through controlled trials to be effective for depressive symptom reduction. Cognitive-behavioral therapy (CBT) (Gallagher-Thompson & Steffen, 1982; Goldapple et al., 2004) and its derivatives dialectical behavioral therapy (DBT) (Linehan, 1993; Lynch, 2000; Lynch et al., 2007; Lynch, Morse, Mendelson, & Robins, 2003), problem-solving therapy (PST) (Arean et al., 1993; Nezu, 1987), and interpersonal psychotherapy (IPT) have been the most widely studied.

Our group of therapists have been trained in IPT and PST and have commented to the author on their view of the differences in approach: PST seeks to quickly get the identified patient to select and commit to completing very specific tasks (problem solving) between weekly session after brainstorming with the PST therapist to weigh all the pros and cons of alternative actions. The hoped-for effect is to disprove patients' own pessimistic or nihilistic attitudes toward themselves and activate their ability to take initiative to solve problems for themselves. If no problem can be identified, then pursuit of pleasurable activity is sought after. The hope is that repeated cycles of problem solving and strict adherence to a model, using the PST worksheet that calls for six to eight 30 min sessions, will activate the patients and combat the demoralization from their depressive illness and that problem-solving successes will continue to generalize to every new problem the patients encounter after the PST concludes. The IPT therapists in our group who subsequently learned PST have come to appreciate the value of highly specific committed actions on the part of the patient as well as the value of repetition. IPT by comparison is described as more open to devote adequate time to hear about the range of the patient's feeling state. IPT was also better able to address complex issues like the myriad of feelings and thoughts that were present in one man whose wife was dying of cancer or when the widest view possible was necessary to asses the input and characteristics of all caregiving parties in an extended family with different agendas. Perhaps one can think of PST as highly focused on a few representative problems, whereas IPT seeks to establish the broadest view possible that includes all contributing parties with a stake in the welfare of the identified patient. PST does not work simultaneously or jointly with caregivers as does IPT-ci. Another distinction is that PST is short-term focused, and IPT-ci takes the long-term view as it anticipates likely changes over time in cognitively impaired patients. The interested reader is referred to the reference list to learn more about CBT, DBT, and PST as this book will focus exclusively on IPT and its adaptation for use in cognitively impaired elders with and without co-occurring depression.

How is the choice typically made for treatment of depression? Often, it is made by the expedient demands placed on the primary care doctor who has precious little time to devote to listening to precipitating problems and who often quickly prescribes a relatively safe trial of antidepressant medication when a threshold of depressive symptoms are detected. If there is an obvious stressor like bereavement or a threatened divorce on the horizon, then "counseling" may be considered as an alternative (or adjunctive treatment) to antidepressant medication. It could be argued that every depressed person needs talking therapy to try to make sense of the experience of depression, to search for possible risk factors, to explore better ways to handle troublesome relationships, and perhaps most importantly to help repair relationships that may have been damaged by the pessimism and irritability that often accompanies depression. Combination treatment makes sense to most people to be able to achieve the most rapid and most complete recovery from depression. Some individuals cannot tolerate the side effects of antidepressant medications or simply refuse to take them, which leaves psychotherapy alone as the only other option as ECT and other neuronal stimulation therapies are reserved for very severe cases and are not available in many locations. Some episodes of depression do remit spontaneously over time but, on the other hand, some individuals have suffered from chronic depression that has lasted for 10 or more years if left untreated.

Other Diagnoses

Bipolar Disorder

Bipolar disorder is defined as recurrent depression with at least one lifetime episode of mania (Bipolar I) or hypomania (Bipolar II). There is stronger evidence for a genetic predisposition for bipolar disorder than unipolar depression, and the standard of care requires the use of mood-stabilizing medications such as lithium, or one of the antiseizure medications such as lamotrigine (Lamictal®), valproic acid (Depakote®), or carbamazepine (Tegretol®). Bipolar disorder usually reveals itself in the second or third decade of life, but milder versions can exist undetected until later in life when other factors (such as an illness or drug that affects brain function) cause an exacerbation of symptoms that finally prompts the appropriate diagnosis.

Co-occuring Anxiety

Anxiety symptoms, such as feeling keyed up, nervous, fidgety, restless, or having difficulty concentrating, which may or may not be accompanied by heart palpitations, sweating, headaches, muscle tension, pacing, or diarrhea, can co-occur with depression, making the afflicted individual doubly miserable. (Driscoll, Karp, Dew, & Reynolds, 2007). Anxiety symptoms can also exist in an individual without depression as in the *DSM-IV-TR* diagnostic categories of generalized anxiety disorder (GAD), panic disorder, phobias, obsessive-compulsive disorder (OCD), or posttraumatic stress disorder. A flare up of one of these conditions can precipitate or exacerbate a depression syndrome and vice versa. Disorders such as GAD can occur, and often do, only in the context of an episode of depression and resolve completely when the depression resolves.

The treatment of depression is made more difficult by the presence of a co-occurring anxiety state (Cairney, Corna, Veldbuizen, Herrmann, & Streimer, 2008; Lenze, 2007). It takes longer to achieve a full recovery, and relapses are more common. Co-occurring anxiety may require the concomitant use of specific antianxiety medications (such as benzodiazpines, alprazolam, and lorazepam) or less commonly atypical neuroleptics (such as quetiapine, olanzapine, or aripiprazole) to treat the excess anxiety simultaneously.

Personality Disorders

Personality style can predispose some individuals to depression or even suicide (Heisel, Links, Conn, vam Reekum, & Flett, 2007). For example, a great deal of research has shown that persons diagnosed with borderline personality disorder who perceive that they have experienced a rejection (such as a romantic breakup or even a therapist going away for a vacation) can become precipitously depressed. When life changes precipitate a state of crisis, for example, when a new diagnosis of an incurable illness is made in an individual with a narcissistic or obsessive-compulsive personality disorder, the impact of the diagnosis forces a confrontation with his or her usual coping ability that cannot always absorb such a reality, and severe depression can result (Pilkonis & Frank, 1988). For example, individuals with narcissistic personality disorder may have a strong need to see themselves as privileged, superior, and entitled (to good health among other things). A person with obsessive-compulsive personality disorder operates on the premise that life's problems can be compensated for thorough stringent control of thought and behavior. When individuals with these two personality disorders are confronted with an incurable illness, their lifelong coping strategy is turned upside down, and they can feel completely overwhelmed and become seriously depressed.

Individuals with narcissistic personality disorder can express rage for being robbed of their health or blame the doctor or the health care system for failing them. When the incurability of the illness sinks into their consciousness, these same individuals can feel hopeless and depressed as they feel like they have lost their self-image of robustness, power, or entitlement. Individuals with obsessive-compulsive personality disorder who have been accustomed to meeting life's challenges by more intensive, obsessive attention to detail or by working harder or longer may become severely depressed when they are confronted by an illness they cannot handle by their usual coping strategies. Similarly, when the person being most relied upon dies in the life of a person with dependent personality disorder, a sense of panic and depression can be precipitated as the person gropes for someone else to replace the lost object of dependency and if no one is available, a desperate sense of demoralization can quickly set in as was seen in the case of Millie in Chapter 1.

Persons with borderline personality disorder who have shown a chronic pattern of poor coping skills, chronically unsatisfying relationships, or who often feel enraged or suicidal are probably best treated with dialectical behavior therapy (DBT), which was specifically designed to help them through the use of intensive and frequent interventions in a group setting along with skill-building classes that teach alternative techniques for better regulation of overwhelmingly strong emotions (Hirschhorn, Lohmueller, Byrne, & Hirschhorn, 2002).

Depression and Cognitive Impairment

When depression and cognitive impairment co-occur, the morbidity conveyed by either affliction can magnify the other (Alexopoulos et al., 2005; Baldwin et al., 2004; Kalayam & Alexopoulos, 1999; Stein, 2008). For example, someone with mild memory loss and somewhat diminished social functioning can show worsened memory and confusion, and can become functionally incapacitated when he or she develops depression as well. With adequate treatment of depression, this same individual's cognitive state will usually improve back to its "pre-depression" level. For clinicians meeting individuals for the first time who show features of both cognitive impairment and depression, it is often impossible to tell how much their cognitive function, particularly the speed with which the brain processes information, will improve until the depression is adequately treated (Butters et al., pp. 587–595). In fact, when trying to evaluate the level of cognitive impairment in a given patient who also shows features of depression, neuropsychologists often conclude that they must wait until the depression, like a malicious fog, is lifted with adequate treatment, to be able to be confident that their test conclusions reflect the person's true cognitive abilities. Adequate treatment for depression is therefore imperative to maximize cognitive ability and functional capacity.

Cognitive impairment can also interfere with depression treatment outcomes (Alexopoulos et al., 1997; Alexopoulos, 2003; Barnes, Alexopoulos, Lopez, Williamson, & Yaffe, 2004; Butters et al., 2004; Gabryelewicz et al., 2004; Lyketsos et al., 2002) through poor compliance with medication regimens such as missed doses or inadvertently doubled doses when the prior consumed dose was forgotten. Increased supervision of medication such as the use of organized medication trays may help to alleviate this risk. Several studies have shown that minimal cognitive impairment (MCI) and depression have independent effects on disability (Alexopoulos et al., 2005; Burdick et al., 2005; Lockwood et al., 2000; Steffens, Hays, & Krishnan, 1999; Tuokko, Morris, & Ebert, 2005). One intriguing finding that has come to light from recent research has been the increased risk for developing cognitive impairment or dementia in those with a history of recurrent depressive episodes. Through brain imaging research, we can now detect shrinkage in the hippocampus in patients who have repeated episodes of depression, an area involved in memory processing and the same area of the brain where the hallmark pathological changes of Alzheimer's disease (microscopic plaques and tangles) show up first (Sheline et al., 2004). These findings give further impetus to treating depression to full remission and preventing recurrences to try to prevent or delay such deleterious brain changes that might hasten the onset of dementia. Many lines of further clarifying research are ongoing at this writing.

The sequencing of medications used for individuals with depression and cognitive impairment usually involves treatment with antidepressants first and subsequently with cholinergic-enhancing agents for the cognitive impairment once the depression has resolved and when the picture of cognitive impairment indicates a dementia.

Collaborative Care

Just as greater success has been achieved in maximizing the health of those with diabetes and asthma with a team approach versus a single-doctor approach (the collaborative care model), the same has been shown for maximizing the treatment of depression for older individuals. As the vast majority of older individuals with depression are seen in primary care medical practices, several systematic attempts to integrate specialty mental healthcare have been successfully implemented such as the PROSPECT (Bruce et al., 2004), IMPACT (Unutzer et al., 2002), and PRISM-E studies (Skultety & Rodriguez, 2008). In these models, mental health specialists (usually social workers or nurses) were available to engage, evaluate, provide psychotherapy, and coordinate good compliance with medication regimens within the same setting in which patients receive their primary medical care. Off-site psychiatric expertise was available in these studies to fine-tune treatment and to consult on the most difficult cases. The on-site availability of these mental health specialists and their visibility and collegiality with primary care doctors resulted in more patients accepting an offer of psychotherapy, as well as affording more time devoted to psychosocial issues or stressors than the patient would have otherwise received in a typical primary care doctor–patient relationship. These collaborative care models achieved greater improvement in depression symptom reduction (particularly suicidal ideation).

The same mental health specialists could be trained to provide collaborative care for patients who show evidence of cognitive impairment (with or without concomitant depression). The IPT-ci model of care delivery would equip these mental health specialists to diagnose and collaboratively treat cognitive impairment, depression, or their combination. Furthermore, the principles of IPT-ci equip the mental health specialist to integrate the caregivers into the treatment plan from the outset, to fully educate them, to assess their competence as caregivers, to enlist their collaboration, to refer them for their own help (regarding social service, legal issues, mental health, or substance abuse) when indicated and to prepare them for the evolving need to adapt to the changing status of a dementing loved one (the identified patient) over time. Further details will be elucidated in Section II. We will now turn our attention to an overview of cognitive impairment, with particular emphasis on executive dysfunction.

5

The Cognitive Impairment Spectrum: MCI to Dementia

In our consideration of cognitive impairment, we will take up the worst forms first by focusing on dementia and delirium. Dementia is essentially chronic brain failure, while delirium is acute brain failure. We will begin with delirium or acute brain failure, which can be characterized by a variety of symptoms such as confusion, mistaken perceptions (e.g., thinking a crack in the plaster is a spider), or wild agitation, but its hallmark is a fluctuating state of attention. A delirious person can fall asleep in midsentence and awake later with great fear or agitation as if he or she is having a nightmare while being awake. A common clinical history is that the afflicted person was "fine" or "his or her old self" a few days before but now shows a variety of possible symptoms, which can include agitation or intense sleepiness, depressed mood, paranoia, visual hallucinations such as seeing spiders or snakes, or motor and speech incoordination. The length of delirium onset is usually hours to days. Delirium can affect any age group but it is seen more commonly among the elderly, who more frequently suffer from serious medical illnesses.

Delirium

The brain is the most sensitive organ to perturbation of the internal status quo of the body. Universal evidence of this is the malaise we all feel when we have a common cold and when we can't seem to concentrate and we would prefer to avoid strenuous mental tasks. Any illness that has toxic effects on the brain can cause delirium if it is severe enough such as kidney, liver, lung, or heart failure. Examples include decreased oxygenation of the brain, a long list of metabolic derangements that commonly include perturbations of the dissolved minerals in the blood (such as potassium, calcium, or sodium being too high or too low). A common cause of delirium in the elderly is the toxic effects of a severe urinary tract infection. Low oxygen, low blood sugar, infections, metabolic disturbances, drug effects, and withdrawal from sedative drugs are all common causes of delirium, which can accompany any severe illness such as pneumonia or a bout of congestive heart failure in a vulnerable individual (see Table 5.1). The presence of delirium can thus be an indicator of a severe illness that is yet to be obvious. For example, mental confusion can be the

Table 5.1 Delirium Clinical Features

- Impaired level of consciousness with reduced ability to direct, sustain, and shift attention.
- Global impairment of cognition with disorientation, impairment of recent memory, and abstract thinking.
- Disturbance in sleep–wake cycle; excessive dreaming with persistence of experience during wakefulness.
- Psychomotor disturbances including agitation or hypoactivity.
- Emotional lability.
- Perceptual distortions, illusions, and hallucinations—characteristically visual.
- Speech may be rambling, incoherent, and thought disordered.
- There may be poorly developed paranoid delusions.
- Most commonly: Onset of clinical features is rapid, with fluctuations in severity over minutes and hours (even back to apparent normality).

Source: Kupfer, D., Horner, M. S., Brent, D., Lewis, D., Reynolds, C., Thase, M., Travis, M., & Horner, M. X. (Eds.). (2008). *Oxford American Handbook of Psychiatry*. New York: Oxford University Press. Reprinted with permission.

first sign of a cardiac arrhythmia such as atrial fibrillation that results in a net decrease in oxygenation to the brain. Delirium is a serious medical condition that results in death in as much as 30% of cases. Delirium can be a medical emergency and should always trigger a prompt medical investigation of the cause, as illustrated in the following case vignette.

CASE VIGNETTE 7: A TEACHER FAILS HER EXAM

A 57-year-old high school teacher is admitted to the hospital for a biopsy of her cervix after she was noted to have an abnormal pap smear. Upon interviewing her, despite no history of mental illness, depression, or cognitive problems, she could not recall the date when she previously met the doctor. On closer questioning, she could not be sure what the current date was and had trouble calculating simple math problems. This sudden change in cognitive ability, especially with a previous history of very high functioning, suggested possible delirium or acute brain failure. A series of screening blood tests were quickly sent off to look for possible causes. The results showed, unfortunately, that she was in acute renal failure due to cervical cancer that had grown large enough internally to compress both of her ureters, the tubes that transport urine from the kidneys to the bladder. This blockage disallowed her kidneys to do their work of removing toxins from the body, particularly

CASE VIGNETTE 7 (*Continued*)

urea, a by-product of protein metabolism. An excessively high concentration of urea in the blood is one cause of delirium. Normal levels of blood urea nitrogen (BUN) are around 20 standard units, but hers was 78. She needed to have emergency hemodialysis that night before they could prepare her for surgery to remove the tumor and the compression that it was causing on her ureters. When her renal failure was thus corrected, her cognitive ability returned to normal.

Older individuals who are already in the early stages of a dementing illness such as Alzheimer's disease are at a higher risk to develop delirium when they get medically ill since they have already lost some brain function due to the dementia, in other words, they have less reserve brain capacity as the following case will illustrate.

CASE VIGNETTE 8: DELIRIUM DUE TO WITHDRAWAL OF SEDATING MEDICATIONS

An 80-year-old widow with mild dementia (probable early Alzheimer's disease) developed pneumonia and was admitted to the hospital and treated with intravenous fluids and antibiotics. Her fever came down, and the x-ray of her lungs showed that her pneumonia was improving within the first 48 hours. She had been pleasant, clear minded, and cooperative throughout her hospital stay. On the fourth day of her hospitalization, however, she became confused, combative, paranoid, and agitated. She thought the nurses were trying to kill her, and she pulled out her intravenous lines and tried to "escape." She needed to be restrained for her own safety and had to be given some medication to calm her agitation in order to continue treatment for her serious pneumonia.

The result of this puzzling change in mental status was solved when her daughter went to her home and searched her bathroom medicine cabinet and found a bottle of alprazolam (Xanax®), which was prescribed by a doctor other than her primary care physician. Her mental symptoms could now be understood as a withdrawal state from sedative medication.

This phenomenon is seen frequently when those who use sedatives or alcohol regularly are admitted to a hospital but their regular use (or addiction) is not recognized. It is commonly seen when someone is admitted after a motor vehicle accident and perhaps undergoes surgery for orthopedic stabilization. The trauma team is so focused on the patient's injuries that the risk of withdrawal is initially overlooked.

These two examples illustrate the potential medical factors that can affect mental functions such as mood stability, alertness, agitation, concentration ability, sleep, and cognitive ability.

The take-home point is that any acute change in mental status requires consultation with a physician or potentially a trip to the emergency room to consider a diagnosis of delirium that can indicate a serious underlying medical problem, medication reaction, or perhaps a withdrawal state if a sedating medication or heavy alcohol usage was abruptly stopped. Because delirium indicates that the brain is already failing to function correctly, the underlying offending condition is either severe or indicative of someone who is frail, has little reserve capacity, or has multiple coexisting problems. Delirium can be an indicator that organ systems are failing, which can be a harbinger of potential death if the condition is not treated immediately as in the case with the teacher with renal failure.

Dementia

Now let us turn our attention back to dementia or chronic brain failure. In contrast to delirium, with dementia you do not see fluctuating levels of attention or consciousness. The hallmark of dementia is memory loss in addition to deteriorated function in one of several additional areas of function such as language impairment (difficulty naming objects or errors in syntax, mispronunciation, or word substitution), visual spatial ability (misrecognizing faces or becoming lost in familiar places), or personality changes (for example, becoming disinhibited by blurting out socially inappropriate criticisms or being self-absorbed, with less attention and care paid to the needs of others). These symptoms need to be severe enough to interfere with social or occupational functioning to merit the diagnosis of dementia.

One type of dementia called frontotemporal dementia usually presents with personality change first or disinhibition, as the following case vignette will illustrate.

CASE VIGNETTE 9: SHOULD HE GET FIRED?

A 63-year-old county policeman was placed on suspension for walking away from his duty when transporting prisoners from jail to the courthouse because he got the impulse to get a cup of coffee. His partner was shocked when he just walked away as if he had no worry of potential consequences. A mental status exam showed his memory, calculation ability, and orientation to be intact. He scored 29/30 on the mini–mental state examination (MMSE). In gathering more history from his wife, however, a pattern of impulsivity, poor

CASE VIGNETTE 9 (*Continued*)

insight, and increasing self-centeredness emerged. She felt that he was acting strange. He would leave the house to take the dog for walks that lasted several hours without mentioning to her where he was going or when he would be back. When she confronted him, he did not seem to be able to see why she would be annoyed with his behavior, and he continued similar behavior in other circumstances. Although an MRI scan was normal, a SPECT scan that maps blood flow to various parts of the brain showed poor perfusion of the frontal lobes of his brain. He was therefore diagnosed with frontotemporal dementia and instead of being fired, he was placed on medical leave (he was unable to perform his job safely due to his dementia) and applied for long-term disability.

The causes of the frontotemporal group dementias are mostly degenerative neurological diseases, the causal triggers of which remain poorly understood, but clear genetic links are evident for some types (Bertram & Tanzi, 2005; Grossman, 2002). This group of dementias is rare compared to Alzheimer's disease (AD), the most common cause of dementia (69%), which is named for the doctor Alois Alzheimer, who first characterized the symptom pattern and noted the characteristic microscopic changes in the brain known as plaques and tangles in the early 1900s.

Alzheimer's Disease

Typically, AD begins with slowly progressive memory loss over a period of years. Complaints of word-finding difficulty, forgetting recent events, and losing objects are common features early on. The memory problem is specific to the laying down of new memories but not the retrieval of old memories such that the individuals have difficulty retaining a list of words or a shopping list, but can often still recall the names of their high school friends (in the early stages).

In the brain of victims of AD, more hallmark microscopic plaques and tangles are present in the areas responsible for processing memory (the temporal lobes) and the disease worsens in proportion to the increasing load of these hallmark changes in the brain. In AD, microscopic cell death is taking place, resulting in measurable shrinkage or atrophy that can be seen on CAT or MRI scans of the brain. There are cases, however, that do not have abnormal-looking brains on these scans even though they have typical AD symptom abnormalities at the cellular level. The cause of AD is not known, but we do know a lot about what is going wrong, just not the initial triggers that begin a cascade of changes within the brain. A lengthy digression on hypothetical mechanisms of AD is beyond the scope of this book, but the reader should know (and this is hopeful news for patients and family members) that intensive research is taking place on multiple fronts to better understand this debilitating disease and to produce effective treatments or, better yet, preventative strategies. For further information, several excellent references are available (Bertram & Tanzi, 2005; Halliday, Robinson, & Shepherd, 2000; Mullan, Crawford, Axelman, & Houlden, 1992; Welsh-Bohmer, Gearing, Saunders, Roses, & Mirra, 1997).

There is ample evidence that AD can be a genetic disease that runs in some families (Winblad & Poritis, 1999). This is usually the early-onset type that can begin to show symptoms in a person as early as age 40. Several genes have been identified that clearly convey increased risk, and these genes have been linked to clear family pedigrees that prove a linkage to the abnormal gene. The AD story is more complicated, however, as the majority of AD cases are still "sporadic," that is, those that occur without a pattern of family inheritance necessarily and have a later onset (in the seventh decade of life on average). Unless there is a clear family history of early-onset AD in first-degree relatives (such as parents, siblings, or children), all that can be said about the risk to family members is that, in general, the more afflicted family members you have, the greater your own risk is of developing AD; however, the risk might skip over you entirely. This answer given to concerned family members is as accurate as we can give at this time and, of course, it should be coupled with a hopeful tone that research may produce breakthrough findings at any time. There will certainly be a Nobel Prize awarded to those who unravel this mysterious and devastating illness.

There is no cure for AD at this time although several lines of investigation are pursuing treatment options. The types of neurons that predominately control memory use a neurotransmitter or chemical messenger called acetylcholine to communicate between each other in complicated networks or circuits of thousands of neurons connecting various areas of the brain that carry out specific functions. The development of drugs that enhance the supply of acetylcholine (cholinergic enhancers) such as Aricept®(Feldman et al., 2001), Exelon®, and Razadyne®(Qaseem et al., 2008; Ringman, 2006) have been shown to improve memory, attention, mood, and engageability in AD sufferers. Some patients show dramatic improvement when these drugs are begun; others show no obvious change. On average, these drugs produce a modest short-term improvement in cognition. The real benefit of these cholinergic enhancers, however, is their ability to slow further deterioration of cognitive impairment. Population studies have been done where some AD victims are randomly assigned to the drug and others to a placebo in double-blind fashion (neither patient nor treating doctor know the assignment), which is the modern standard way of testing for a true beneficial effect. In these studies, a clear benefit for slowing the downward progression of AD severity was found in those who were randomly assigned to active drug versus placebo (Kaufer, 2002; Lanctot et al., 2003; Lyness, 2002). These drugs can potentially delay the need for nursing home placement by 2 years. When we consider the increasing number of baby boomers aging into their 70s and 80s, this difference adds up to a huge cost savings in healthcare dollars in addition to the direct benefits to the individual patient.

Assessing whether these cholinergic-enhancing drugs slow the rate of further deterioration cannot, of course, be determined in the same individuals as they are either on the drug or not. To determine whether these drugs indeed slow progression, an experiment had to be devised to compare one group who got the drug to another who did not.

In the studies mentioned earlier, if the group on active drug is then switched to placebo after a year, the rate of deterioration in cognition accelerates such that they "catch up" with the group that was on placebo all along. This deterioration is not recoverable with reinstating the drug subsequently. The current standard of care is thus to begin cholinergic-enhancer medication as soon as a diagnosis of dementia is made to achieve maximum benefit in slowing the progression over time. At this writing, only donepezil (Aricept®) is FDA approved for mild dementia.

There is still debate in some circles about the cost–benefit ratio of these drugs; for example, in Britain, the National Health Service has ruled that the modest benefit does not justify the cost (Burns & O'Brien, 2006). In the United States, the prevalent perspective is that any measure that can slow this devastating illness should be taken. Medical economists point out that, taken over an average lifespan, this beneficial difference that could delay nursing home placement by a year or two provides a cost saving when you compare the cost of drug treatment to the cost of nursing home care.

It is important to note that since cell death is occurring in AD, these cholinergic-enhancing drugs do not repair the damage or even arrest the underlying disease process (which we don't understand completely). Therefore, Alzheimer's sufferers still inexorably march toward increased severity and eventual death in the late stages, where speech, motor function, and all cognitive processes can be grossly impaired. Death usually follows from pneumonia in this weakened state unless the person dies from another ailment first. The cholinergic-enhancing drugs Aricept® (donepezil), Exelon® (galantamine), and Razadyne® (rirastigmine) merely improve the functioning of the acetylcholine-using neurons that are weakened by AD but are not yet dead (Kaufer, 2002; Lyness, 2002).

Another drug called Namenda® works through a different mechanism by blocking the toxic effects of another neurotransmitter called glutamate, which can cause damage by being excessively excitatory. This type of drug has been shown to have neuroprotective effects and also to slow the progression of AD; thus it is commonly used in conjunction with the cholinergic-enhancing drugs. Taking two drugs daily, of course, increases the cost of the overall treatment but this regimen has become the current best practice (Cummings, Schneider, Tariot, & Graham, 2006). As nausea is a common side effect of the cholinergic enhancers, some patients cannot tolerate them, and in which case using Namenda® alone is the only option.

Vascular Changes in the Brain

The second most common cause of dementia (about 15–20% of cases) is due to vascular changes in the brain. The accumulation of enough tiny strokes due to blocked arteries eventually leads to a loss of brain power as the underlying brain circuits are disrupted. In explaining this to family members, the author sometimes uses the metaphor of asking them to imagine a telephone switching station for a medium-sized town where you can see bundles of wires organized by district for routing phone calls. Now imagine that cutting small bundles of wires with a wire cutter is like having a small stroke that damages a small collection of neurons. If you imagine the wires to be like neurons in the brain that carry communication signals, then cutting a few wires randomly would affect phone service for a few houses, but overall the phone system remains intact. If you instead make 1,000 random wire cuts, then a picture begins to emerge of a failing phone system (or brain that has accumulated thousands of tiny strokes).

The risk factors for these cerebrovascular "events" are the same as those for atherlosclerotic heart disease: smoking, high cholesterol, diabetes, high blood pressure, a family history of strokes or heart attacks, and a heart arrhythmia known as atrial fibrillation (Miller et al., 2002). The pattern of cognitive impairment symptoms is more variable in vascular dementia than in AD and often progresses in stepwise fashion with more drastic drops in function after a new round of small strokes occur (which the person may not even know he or she is having). There is a certain randomness to the brain areas affected by these tiny strokes such that symptom patterns reflecting the underlying brain damage can vary accordingly. Once the brain tissue fed by a specific artery is choked off by atherosclerotic disease, a blood clot can form on the damaged arterial wall and can cause a complete occlusion of that artery and that brain tissue dies and is gone forever. In the first hours of an evolving stroke or clot in a major artery, however, administering a "clot busting" drug might reopen the blockage and restore blood flow and rescue some brain tissue function. This is now commonly done in the emergency room after a stroke has been diagnosed unless there are other reasons (such as an ulcer, for example) not to give the "clot busting" drug, which also has the side effect of increasing the risk of bleeding elsewhere.

In vascular dementia, in which there is an accumulation of tiny (microscopic) strokes, we more commonly see personality changes such as apathy or disinhibition than we do in the early stages of AD. Also more common are related symptoms such as motor weakness or uncoordinated walking, which are due to tiny strokes that are affecting areas of the brain responsible for controlling motor movements. In contrast, the typical microscopic changes of AD (plaques and tangles) occur preferentially in the deeper structures of the brain (hippocampus and entorhinal cortex) and spare the motor control areas early on.

The strategy for treating vascular dementia is to try to prevent any more ministrokes from occurring by using anticoagulant medications such as dicoumoral (Coumadin®), dipyramadole (Plavix®), or aspirin to decrease the tendency of the blood to form clots that deposit themselves on the damaged arterial wall and cause more strokes. Other protective strategies include better control of blood pressure (as high pressure can lead to damage to arterial walls), treating any underlying heart disease that increases the risk of forming clots (such as atrial fibrillation), maximizing the treatment of diabetes and high cholesterol, and quitting smoking. All of these are risk factors for forming atherosclerosis. The cholinergic-enhancing drugs, such as donepezil (Aricept®), have also shown some benefit in treating vascular depression by improving the functioning of neurons that are acetylcholine containing that were damaged by the strokes (Kaufer, 2002; Lyness, 2002).

Other Common Causes of Dementia

Less common causes of dementia include a range of neurodegenerative diseases, such as multiple sclerosis, Huntington's disease, Parkinson's disease, Lewy body dementia, and Pick's disease; the latter is one cause of frontotemporal dementia as described in the case vignette of the errant policeman (case vignette 9). Lewy body dementia derives its name from characteristic microscopic lesions seen in the brain known as Lewy bodies (named after their discoverer). This dementia often shares features of Parkinson's disease such as slowed muscle movements, rigidity, or tremor. The mental symptoms of Lewy body dementia, in addition to memory loss and poor insight, also commonly include vivid visual hallucinations, such as seeing people or animals. These hallucinations are often not frightening but can be perplexing. These patients will say things like "I don't know why those little girls come and sit on my couch every afternoon, but they don't bother anyone." The vividness of these visual hallucinations makes them seem very real, and the patient's decline in insight seems to suspend his or her critical faculties, such that the patient takes them at face value despite illogical inconsistencies obvious to all cognitively intact bystanders. Antipsychotic medications can help reduce the hallucinations, but these patients are often very sensitive to these drugs, and thus they need to be dosed very carefully.

Infectious diseases like syphilis and AIDS can also cause dementia. Metabolic derangements such as from low sodium or vitamin B12 deficiency are usually temporary conditions that cause reversible symptoms if caught early enough. However, if a deficiency state goes unrecognized for a long period of time, it can cause permanent brain damage too, as the following case vignette will illustrate.

CASE VIGNETTE 10: HUSBAND THREATENS WIFE WITH KITCHEN KNIFE

Wilbur, a 79-year-old man, was court-committed to a psychiatric inpatient unit after a 3-month pattern of increasingly hostile behavior toward his wife, which culminated in his threatening her with a kitchen knife in hand. His screening blood work revealed a very low B12 level, and replacement therapy was begun by intramuscular injections. His mental status showed poor recall ability, limited insight, paranoid thinking, and a tendency to become fearful, angry, and impulsive.

Usual causes of B12 deficiency are poor intake or poor absorption. A protein called intrinsic factor that is secreted in the stomach is required to bind to B12 molecules before this complex can be absorbed in the last section of the small bowel (the terminal ileum). Anything that interferes with these steps can lead to a deficiency state such as chronic diarrhea from a malabsorption syndrome, lack of intrinsic factor from malfunctioning gastric (parietal) cells, or a missing terminal ilium (from surgery, for example). The latter was Wilbur's problem, as the last four feet of his small bowel had been surgically removed in his 20s following a stab wound during a bar fight.

Apparently at the time of his surgery, no one recognized that he would no longer absorb vitamin B12, and he was probably chronically low for the rest of his adult life. This chronic deficiency appeared to have induced permanent brain damage (dementia) as vigorous replacement of B12 by injection did not improve his symptoms. Other psychotropic drugs were required to try to decrease his impulsivity and render him safer to rejoin his wife. His wife also needed a great deal of psychoeducation to understand his condition and was instructed to seek help immediately if he showed further signs of hostility or impulsive aggression.

Mixed Causes and the Prodromal Period

Recent autopsy studies of demented individuals show a surprising number with a mixture of brain pathology, when studied under a microscope, such as a combination of AD changes as well as multiple small strokes (Sweet et al., 2004). These findings suggest that "pure" causes of dementia may be less common and "mixed" causes may be more common. Treating these conditions therefore requires a thorough evaluation of all risk factors and appropriate treatment for each one.

Recognizing the prodromal period of symptoms that eventually leads to dementia is a topic of keen interest. Since all of these diseases cause cell death, it is recognized that any effective treatment must try to intervene in the earliest stages to limit or prevent permanent damage at the cellular level that cannot be reversed. It is known that the brain damage characteristic of AD (the plaques and tangles) begins to accumulate for 10–15 years before symptoms begin to show. It seems that a threshold of damage needs to occur in the brain—which has a certain amount of excess or reserve capacity—before certain brain circuits, like those controlling memory, begin to break down and function poorly. If we could somehow identify those individuals who were destined to suffer from AD, we could search for preventive treatments. The search for biomarkers that would reliably predict AD is a hot topic of research that is currently exploring genetic markers, spinal fluid proteins, blood components, and sophisticated brain imaging. To date, none has produced a practical, reliable, predictive test. Testing for cognitive changes using sophisticated neuropsychological test batteries can now identify those who have the characteristics of "probable" AD with about 90% accuracy when these results are compared to eventual autopsy results.

Even if reliable biomarkers were available, it should also be stated that many people worry about the impact of the stigma of being identified as having a risk for an illness before it is evident. There is the additional worry that insurance companies would use the information to deny coverage as it would be a predictor of future healthcare expenses.

Minimal Cognitive Impairment

Individuals who show some definite signs of cognitive impairment in either memory, language, insight, visual spatial skills, motor coordination, a change in personality, or declining ability to handle complex tasks, but who do not meet the severity criteria for the symptoms causing significant impairment in social or occupational functioning are said to have minimal cognitive impairment (MCI) (Jefferson et al., 2008; Palmer et al., 2007; Palmer, Backman, Winblad, & Fratiglioni, 2008; Petersen et al., 1999; Rozzini, Chilovi, Trabucchi, & Padovani, 2008;). In other words, their cognitive function is not normal but not impaired enough to cause significant decline in functioning to merit a diagnosis of dementia at the time of testing. There is currently a great debate about what definition or subcategory of MCI indicates a given pattern of cognitive impairment that will predict who will progress to a dementia diagnosis. At this writing, it appears that those who have predominant memory impairment features—the so-called MCI-amnestic type—will convert to diagnosable AD at a rate of about 15% per year on average (Whitwell et al., 2008). It should be noted that the mini–mental status exam (MMSE; Folstein, Folstein, & McHugh, 1975) is a screening tool for possible cognitive decline that has become a standard screening instrument in primary care settings; however, highly intelligent individuals can be experiencing a significant decline in intellectual capacity and still get a perfect score, therefore the MMSE should not be relied upon as the sole measure of cognitive function. Neuropsychologists use a more lengthy and sophisticated series of tests to assess all areas of brain function to get a more refined "big picture" and then use repeated testing over time (most commonly annual testing) to judge whether any further decrement in cognitive function has taken place. Self-awareness of a decline in cognitive ability accompanied by depressive symptoms is also commonly seen (Bruce et al., 2008).

Within the MCI category, subtle or sometimes surprisingly blatant changes in demeanor or problem-solving ability can be seen that constitutes executive dysfunction, a topic we will take up in detail in the next chapter.

When assessing a person with possible cognitive impairment, since there is no definitive test for the most common culprit, AD, doctors embark on a process of elimination to "rule out" a metabolic or infectious cause. Blood tests are performed to look for electrolyte disturbances such as sodium, potassium, or calcium imbalance, white blood cell counts are measured for evidence for infection or inflammation, and special tests are conducted for syphilis or other suspected infections. Screening blood tests for B12 deficiency and hypothyroidism are important as these conditions are easily remediable by restoring normal levels using identical synthetic versions of either vitamin B12 or thyroid hormone. Vitamin B12 is found naturally in dark green leafy vegetables and in organ meats such as liver; however, the most common reason for B12 deficiency is not inadequate ingestion but poor absorption in the small intestine. Taking high doses of vitamin B12 orally may force more into the bloodstream. Alternatively, injections of vitamin B12 get absorbed from the muscle right into the bloodstream bypassing the intestines all together. Both methods have been used for successful replacement, which is confirmed by a repeat blood level within the range of normal. Unfortunately for the case vignette of Wilbur (case vignette 10), it was a case of too little too late.

Chronic Alcoholism and Cognitive Impairment

Chronic alcoholism has many toxic effects on the body, brain, and mind including chronic dementia. Chronic alcoholics are notorious for showing signs of poor coordination, impaired reflexes, and disturbances of normal sleep patterns. As alcoholics usually eat poorly when they consume a lot of calories from alcohol, another B vitamin (vitamin B1) called thiamine can become deficient such that key thiamine-dependent enzymes required for neuronal function stop working. Thiamine deficiency can cause Wernicke–Korsakoff syndrome, wherein victims show characteristic neurological and psychiatric symptoms, which can include tremors, impaired extraocular eye movements, postural instability, incoordination, severe short-term memory loss, and confabulation. Confabulation is a curious phenomenon where patients make up events that seem plausible to them at the moment but show flawed logic. For example, they might state that they met you before, and if you play along looking puzzled, they can weave an entire story of how you were connected with them in some way or other when in fact, you have never met before. These symptoms can usually be reversed by administering thiamine in high dose orally or sometimes intravenously.

Chronic alcoholism can result in a dementia syndrome by itself and is usually characterized by prominent short-term memory deficits. If you look at the brain of someone who died of chronic alcoholism, you can often see degeneration or atrophy in several areas pertaining to memory processing. If a person has been a chronic alcoholic, it also does not preclude him or her from developing AD, however. Thus multiple contributing factors may be determining the final clinical picture.

Use of Brain Imaging and Neuropsychological Testing

The question of whether a CAT or MRI scan of the brain is indicated in a dementia work-up usually depends upon whether there are any subtle indicators such as walking or speech incoordination that justify a search for vascular changes, tumors, or other brain abnormalities (so called "soft" neurologic signs). In the typical symptom pattern of AD (slowly progressive short-term memory loss over a period of months or years as the predominant presentation), one can see some shrinkage or atrophy of the brain in certain areas, but scans of some people can show atrophy and still have no symptoms and still others show no atrophy and have significant symptoms, and therefore these scans are not relied upon as the sole diagnostic evidence by themselves. More commonly, imaging data are used to search for other things like strokes or brain tumors or to quantify the extent of suspected vascular lesions. A SPECT scan as mentioned in the vignette of the policeman who walked off the job (case vignette 9) can reveal blood flow deficits to the frontal lobes of the brain, and it has become the diagnostic confirmatory test for frontotemporal dementia. The yield for diagnosing a treatable cause of dementia using imaging to screen everyone with cognitive changes is low and the scans are expensive. Nevertheless, some form of brain imaging is becoming a standard part of a dementia work up in many settings.

Neuropsychological testing can help differentiate between types of dementia that might have different treatments (e.g., vascular dementia versus Alzheimer's disease), and these tests give the most detailed picture of abilities that are preserved versus impaired. The reason that patterns of impairment vary within a group of victims of AD or vascular dementia is due to the underlying pathological process on the microscopic level having a certain randomness with regard to which brain areas sustain more or less damage (as evidenced by the number of plaques and tangles or small strokes).

Depression and Dementia

Depression can be present in up to 50% of AD victims due to the direct biological effects of the disruption of brain circuits or networks that maintain mood (Lockwood et al., 2000). The cell death that is occurring in AD also affects all neurotransmitters, not just acetylcholine (which is linked to the cognitive deficits of AD). Neurotransmitters known to be linked to mood such as serotonin, norepinephrine, and dopamine can also be reduced, affecting the neural networks they operate within. Those with a prior history of depression earlier in life are, of course, not immune from developing dementia, wherein the clinical picture will reflect both illnesses simultaneously. In addition, in the early stages of AD, self-awareness of declining brain power can also be very upsetting or depressing to the person experiencing it.

Summary

In summary, a dementia "work-up" should be carried out to search for treatable causes of dementia using blood tests, a good overall history, and physical examination. Imaging and the use of neuropsychological testing are tools that can help delineate the extent or a pattern of the cognitive deficits as clearly as possible and the resulting pattern can help differentiate the type of dementia. Once a diagnosis is concluded, afflicted individuals and family members will have many questions about the implications of the findings. A hopeful but not unrealistic attitude is warranted as much has been learned about managing these diseases. Once a probable diagnosis has been established and optimal medical treatment is in place, the focus then shifts to how to help victims and their concerned families cope optimally with the reality of their care and how to reasonably prepare for likely future declines in cognitive function over time. Issues regarding the management of cognitive impairment, depression, and their combination using a modification of interpersonal psychotherapy (i.e., IPT-ci) will be addressed in great detail in Section II.

6

Manifestations of Executive Dysfunction

Executive dysfunction (ED) or executive impairment (to be distinguished from the other ED—erectile dysfunction) refers to deficits in a collection of abilities that are often taken for granted as "normal function." To understand where the term "executive function" comes from, imagine a high-functioning chief executive officer (CEO) of a successful shoe company who is keenly aware of the "big picture" of how his or her company is performing by being knowledgeable about the myriad of details that must be tracked for an efficient and profitable operation. Such an executive would also possesses a keen ability to dissect problems and weigh the pros and cons of potential solutions in order to take decisive action and then to evaluate the results of those actions to keep shoe production humming with flawless precision in order to maximize profits for the company and its shareholders. He or she must track quality control, keep abreast of style changes, continuously fine-tune a marketing plan to maximize sales and profits, and keep an eye on the rising cost of materials, and so on. Our shoe executive must hire employees to do various tasks and must fairly determine competitive salaries that still allow for enough profit to satisfy shareholders who own stock in the company. The executive also needs to monitor employee's productivity, complaints, absenteeism, health problems, and work-related injuries. Many of these tasks might be delegated to managers in a large operation, but managers must also be evaluated for how well they perform.

In a similar way, each of us perform our own daily activities like we have our own internal "executive." We are constantly collecting data, formulating analyses, solving problems, and making myriads of decisions that we carry out so automatically that we often take them for granted, everything from what clothes we choose, where we drive, how we problem solve at work and at home, when to get preventive dental checkups, and so on. are handled by our internal "executive." These capabilities can be observed to begin developing in newborns, who start to take in the world and organize it, learn communication and motor skills, and so on. Young children learn remarkably fast which is a testament to their rapidly expanding brain power. We all have witnessed this seamless development and rapid mastery in children (and in ourselves) such that we take the amazingly complex underlying brain capabilities for granted. It is only when brain injury occurs and there is a breakdown do we appreciate how much complex and seamless coordination is taking place in our various brain circuits. If such a breakdown or inefficiency is modest, resulting symptoms may go unnoticed to the untrained eye.

Signs of Executive Dysfunction

When an individual begins to show signs of cognitive impairment, it does not always begin with memory impairment. Instead, cognitive impairment can present with waning insight or judgment (such as in case vignette 9 of the policeman who walked away from the prisoner he was guarding), more difficulty solving problems than usual, or worsening social graces such as expressing tactless comments or being "blind" to the consequences of one's actions or words upon others. Individuals with ED can also be quick to anger, demanding, impatient, blunt, unkind, or even impulsively violent.

Impaired insight often means that the afflicted person does not realize he or she is perceived by others as changed or having a problem. When attempts are made to point out changes or problems, individuals suffering from ED sometimes react with anger or counterclaim that everyone else except them has a problem. A decline in problem-solving skills might reflect an inability to use a roadmap, trouble operating appliances correctly, or high frustration levels with what most people would consider "everyday problems of living." These individuals might make statements that too many demands are being placed on them (when, in fact, they have not changed appreciably). ED can also manifest as increased self-centeredness in which their awareness of the agendas of others around them declines since this awareness requires abstract thinking capabilities or the ability to "step into someone else's shoes" temporarily to see the world from another's viewpoint. Another common feature of ED is lack of initiative, planning and organizing skills which can lead to not paying bills on time, neglecting seasonal maintenance chores, or the apparent preference to just stay home or passively follow the lead of others without producing any plans or opinions of their own.

Causes of Executive Dysfunction

Executive dysfunction can result from a long list of problems in the brain. Most commonly, ED is caused by the same underlying pathology that causes memory impairment in Alzheimer's disease or vascular dementia but predominantly in "problem-solving circuits" rather than in "memory circuits." Brain damage is occurring on the cellular level in both of these brain disease states with a net loss of functioning neurons. This results in the impaired functioning of circuits that control complex information processing and behavior.

Let's expand on the last point for a minute by examining the relationship of a married couple whom we will call George, who is beginning to show signs of ED, and Linda, his cognitively intact wife.

For George to be aware of how he is perceived by Linda, he must possess the capacity to "imagine" what Linda is thinking about him as if he were watching and listening to himself through her eyes and ears. As mentioned earlier, the capacity for empathy requires the ability for abstract thinking, one of the most complex and higher-order forms of thought, which requires that a number of brain areas talk to each other. For example, George's brain function must weigh subtleties of meaning or judge between options that are similar but not the same, like the difference between two shades of red in a color palette or the difference between being annoyed, angry, and enraged. Individuals who have sustained damage to the circuits or connections between the areas of the brain that need to talk to each other to make these subtle distinctions can begin to show the aforementioned signs of ED. Not surprisingly, the cognitive abilities that require the most brain processing power—such as complex problem solving, abstract thinking, foresight, and planning—are often the first to be affected by an underlying degenerative brain process. These abilities require lots of different brain areas to do their part before an accurate final conclusion or decision can be made. If George's usual routine does not require the use of these highest order functions, initial deficits may go unnoticed by Linda until a more profound or obvious deficit is revealed by an incident such as George becoming lost in a shopping mall or forgetting to open the garage door before backing up the car. These types of seminal events that show that "something obviously went wrong" are what typically cause family members like Linda to bring the individual in for an evaluation with a mental health specialist. Concerned family members are forced to take notice quickly in these instances. The same concerned family, friends, or coworkers might have noticed more subtle changes such as when George began to decline invitations that required new learning (e.g., meeting new people) or lower frustration tolerance (e.g., "there are too many steps" in planning a vacation). Other evidence of change in George can be seen by Linda as "crossing the line" with overly critical remarks or off-color jokes, and so on. These kinds of changes may be understandably attributed to many other things such as overwork, high stress levels, bereavement, inadequate rest, too much alcohol, the need for a vacation, and so on. Individuals in whom Alzheimer's disease is beginning to take hold can show predominant features of ED early on instead of memory loss if the random pattern of damage happens by chance to occur predominantly in information-processing circuits rather than in memory-processing circuits. Since ED is not always obvious to the untrained eye, it is often overlooked or mistakenly attributed by family members and others to stubbornness, laziness, or just plain meanness. Individuals with ED may provoke emotions that can cause others to react punitively or to become fearful, angry, or avoidant, which can further compound the problem. Linda might ponder, for example, whether George has fallen out of love with her, and she might question him critically along these lines.

Rating Scales for Executive Dysfunction

Various rating scales have been used to try to rate or quantify the degree of ED, such as the Executive Interview (EXIT; Royall, Mahurin, & Gray, 1992) or the screening questionnaire the Montreal Cognitive Assessment (MoCA; Nasreddine et al., 2005) which is a one-page test (see Figure 6.1) that covers several areas. See appendix for instructions for administering the test. The dot connecting task (item 1) is a measure of multitasking ability. The MoCA also tests visual-spatial drawing skill and organizational skills with items 2 and 3. In addition to items that test naming, memory, verbal fluency, and calculation ability, the sixth item tests the often impaired ability to inhibit tapping when the incorrect letter is presented. This impaired ability to inhibit the tendency to perseverate when a pattern is established is a key symptom of executive impairment and it can be an indicator of risk for acting impulsively in other situations (e.g., developing "road rage" when there is a perception of being treated disrespectfully). The EXIT scale and others like it are used to quantify ED, to gauge its severity, and to track how much it might improve or decline further over time.

Presentation of Symptoms

The presentation of the symptoms of ED can be sporadic or unpredictable, however, with periods of apparent normalcy in between. Sometimes increased stress can precipitate worsening ED symptoms such as traveling to an unfamiliar place or being placed in an environment where too much is happening simultaneously. Alcohol also lessens inhibitions in cognitively intact people which has an effect (often sought after) to decrease social shyness. In individuals with executive dysfunction, however, alcohol or other mind-altering drugs can lessen inhibitions to the point of abject social impropriety, which the impaired individual may not even be aware of. Individuals with ED in an environment that they perceive to be chaotic or overwhelming can reach their limit of frustration tolerance and blurt out an angry statement that might reflect their own feeling state of not having their needs met, yet their behavior may be perceived by others as "out of line," "over the top," or highly inappropriate.

In the previous chapter, case vignette 9 of a policeman who walked away from the shackled prisoner he was transporting is a good example of ED. The following case vignette is a more typical presentation.

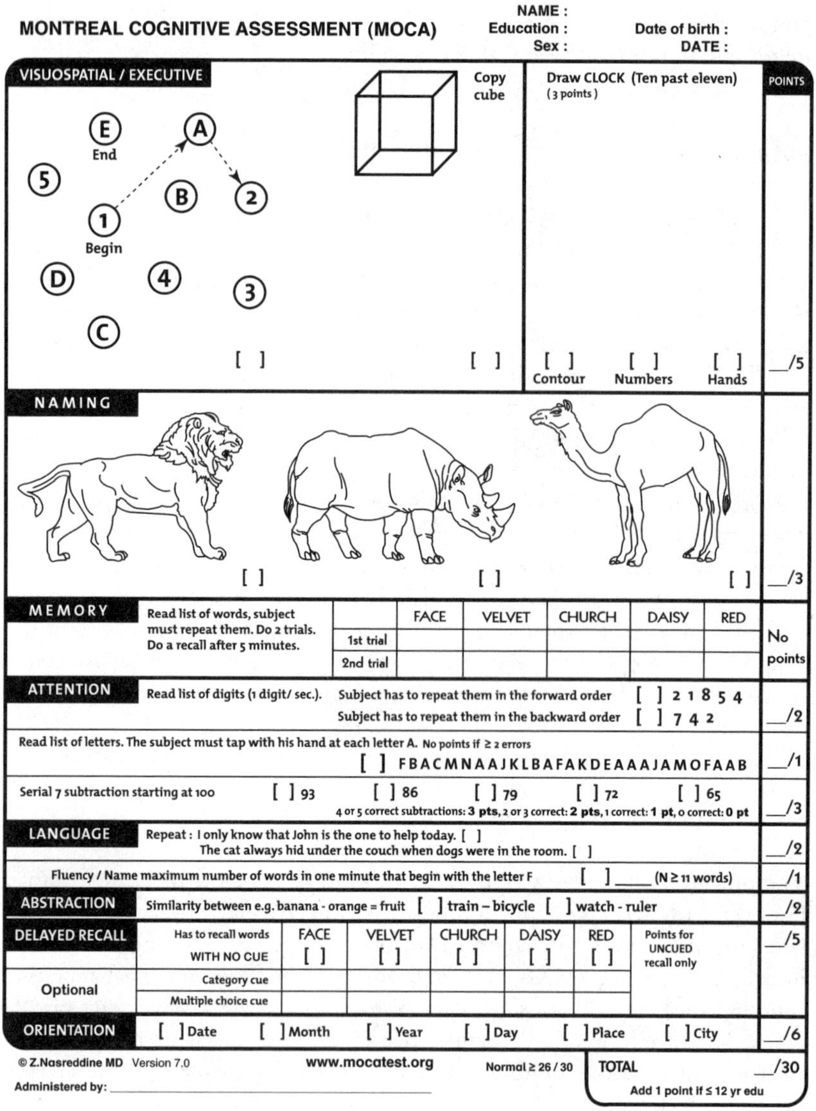

MONTREAL COGNITIVE ASSESSMENT (MOCA)

NAME :
Education :
Sex :
Date of birth :
DATE :

VISUOSPATIAL / EXECUTIVE

[ ]

Copy cube [ ]

Draw CLOCK (Ten past eleven) (3 points)

[ ] Contour [ ] Numbers [ ] Hands

POINTS __/5

NAMING

[ ] [ ] [ ] __/3

MEMORY	Read list of words, subject must repeat them. Do 2 trials. Do a recall after 5 minutes.		FACE	VELVET	CHURCH	DAISY	RED	No points
		1st trial						
		2nd trial						

ATTENTION Read list of digits (1 digit/ sec.). Subject has to repeat them in the forward order [ ] 2 1 8 5 4
Subject has to repeat them in the backward order [ ] 7 4 2 __/2

Read list of letters. The subject must tap with his hand at each letter A. No points if ≥ 2 errors
[ ] FBACMNAAJKLBAFAKDEAAAJAMOFAAB __/1

Serial 7 subtraction starting at 100 [ ] 93 [ ] 86 [ ] 79 [ ] 72 [ ] 65
4 or 5 correct subtractions: **3 pts**, 2 or 3 correct: **2 pts**, 1 correct: **1 pt**, 0 correct: **0 pt** __/3

LANGUAGE Repeat : I only know that John is the one to help today. [ ]
The cat always hid under the couch when dogs were in the room. [ ] __/2

Fluency / Name maximum number of words in one minute that begin with the letter F [ ] ____ (N ≥ 11 words) __/1

ABSTRACTION Similarity between e.g. banana - orange = fruit [ ] train – bicycle [ ] watch - ruler __/2

DELAYED RECALL	Has to recall words WITH NO CUE	FACE [ ]	VELVET [ ]	CHURCH [ ]	DAISY [ ]	RED [ ]	Points for UNCUED recall only	__/5
Optional	Category cue							
	Multiple choice cue							

ORIENTATION [ ] Date [ ] Month [ ] Year [ ] Day [ ] Place [ ] City __/6

© Z.Nasreddine MD Version 7.0 www.mocatest.org Normal ≥ 26 / 30 TOTAL __/30
Administered by: ____ Add 1 point if ≤ 12 yr edu

Figure 6.1 The Montreal Cognitive Assessment Test

CASE VIGNETTE 11A: SPORADIC ODD BEHAVIORS EVOLVE INTO AGGRESSION

The identified patient, Oscar Johnson, is a 79-year-old retired retailer who was referred by his primary care physician to a Geriatric Care Center for help with his depression. His initial presentation appeared to be categorized as a role transition in traditional IPT and was treated as such. He was noted to be withdrawing from his usual activities and was increasingly negativistic and irritable. Oscar agreed to contract for treatment of moderate depressive symptoms and did so alone. He drove himself to his appointments. He presented himself in a matter-of-fact way, but was well dressed and groomed, polite, and articulate. He did not complain of memory disturbance, rather, his chief complaint seemed to be ambivalence about how to pull back from some of his perceived duties that he no longer felt he could keep up with. He was referring to the perceived demands of two older spinster sisters who lived together nearby for whom he had run errands and done odd jobs for his entire adult life. Oscar had retired a year earlier from self-employment as a corner grocery store owner, and he was now considering stopping his part-time job as a counter clerk in a local deli as well. When asked for the reason behind his decision to retire fully, he stated that "everyone makes me mad," "the times have changed," and "young people don't have the same respect we were brought up with." One of his sisters had called him recently asking him to come and look in to a leaking faucet. He seemed to be obsessing about whether to help his aged sisters one more time or to draw the line and tell them he was too old to play this role for them any longer. He wondered out loud why they couldn't call someone else who was younger to help them.

During that initial evaluation, Oscar scored 26/30 on the Folstein Mini–Mental Status Examination (MMSE). As he was retiring and wanted to scale back his responsibilities as the apparent driving force behind his depressive symptoms, the problem area of role transition in traditional IPT seemed appropriate. The interpersonal inventory revealed that his wife was still working and that he had several grown children and grandchildren but did not indicate any relationships that might suggest a role dispute. His two sisters were not aware of his struggle regarding his ambivalence about continuing to help them and therefore his problem was formulated as a role transition and not a role dispute (he was also considering quitting his job in the deli that he also found to be frustrating, another role transition).

Oscar met criteria for major depression and was also treated with antidepressant medication in combination with IPT. With regular weekly visits in IPT, Oscar explored the pros and cons of announcing to his sisters that he no longer wanted to be their ombudsman, and he did manage to say so eventually with the encouragement of his IPT therapist. In subsequent sessions, he processed their reaction, which was overall one of understanding. His sisters thanked him for his years of service and said they could recognize that he was "getting up in years too." They vowed to ask their nephew for any help they might need in the future. Oscar also ended his sporadic hours at the deli and expressed some relief that he was now completely retired and could relax and enjoy it. He subsequently seemed content to do very little throughout the day and seemed to have few goals or ambitions for things he wanted to do during his retirement. His depressive symptoms declined into the range of normal,

CASE VIGNETTE 11A (*Continued*)

he completed his course of IPT, and he agreed to participate in a study that evaluated and followed his cognitive function over time.

His neuropsychological test battery showed him to have reasonably good verbal learning skills, mild visual-spatial difficulties, and moderate impairment of ED. The latter seemed to explain his easy frustration and the difficulty he seemed to have in problem solving (such as coming to a conclusion about whether and how to withdraw his help for his sisters). At this point, his neuropsychological test battery results were categorized as having minimal cognitive impairment (MCI)–nonamnestic type. This meant that he had MCI that was not primarily memory related.

Over the following 2 years of maintenance sessions and with new information introduced by Oscar's wife, a picture began to emerge of more severe impulse dyscontrol, poor insight, poor problem solving, and poor judgment. Repeat neuropsych testing showed that Oscar's cognitive function had deteriorated such that he now met criteria for dementia with a frontal lobe pattern of predominant personality change and disinhibition. A SPECT scan revealed a pattern of blood flow deficits to the frontal lobes that confirmed the diagnosis of frontotemporal dementia, a class of dementia distinct from Alzheimer's disease but degenerative nonetheless as described in Chapter 4. Oscar had been on a serotonin reuptake inhibitor antidepressant known to have anti-agitation and anti-disinhibition properties, but would at times on his own stop taking it (also evidence of poor insight, poor planning, and impaired logic). Therefore, he was switched to a longer acting version of the antidepressant that would continue to provide some protection against severely disinhibited behavior even if he missed a dose or two. He was covetous of his medications and refused to let his wife help him keep them organized.

CASE VIGNETTE 11B: SPOUSE BECOMING A CAREGIVER

Oscar's wife worked as a clerk during the day, and she took off work to accompany him on several visits. She described a series of events regarding Oscar's problem behaviors that caused her a great deal of distress such as social inappropriateness at his own birthday party when he complained that one of his children did not bring him a gift. At other times, his wife reported that he was rigid, dogmatic, and demanding that she agree with his way of thinking, and she reported that he got enraged when she did not. Mrs. Johnson did not know what to attribute this uncharacteristic behavior to. She suspected the effects of idleness from his retirement were the root cause. She found herself feeling anxious and troubled and admitted that she sometimes felt reluctant to come home from work. On one occasion after an argument "over nothing" in a hotel room balcony while attending a wedding out of town, she reported that Oscar locked her out on the balcony for an hour when he became enraged. At another time, he seemed to get so angry that she was afraid he might strike her which had never happened before throughout their entire marriage. She hadn't told any of her children about these incidents for fear that it would lead them to "lower their respect for their father."

One frequent theme for their arguments revolved around one of their daughters who was going through divorce proceedings. Even though no mention of parental financial support was ever made among family members, Mrs. Johnson reported that Oscar became convinced that he and his wife would be expected to provide financial support for their daughter and their grandchildren, which would bankrupt their savings for retirement. This "theme" was raised often by Oscar and seemed to be unalterable by any counter-arguments his wife might try to make.

Taken together, these behaviors suggested a pattern of ED with an inability to see situations from more than one angle regarding his daughter's divorce, difficulty following details or social complexity, and highly personalized interpretations of events with distorted, irrational fears stemming from an inability to sift through options and rank the relative strengths and weaknesses of each of them.

In a series of individual session with Mrs. Johnson separately, contingencies for safety were undertaken in case further aggressive behavior continued. She repeatedly said she did not live in fear of him as these threatening events were only occasional and that he seemed apologetic afterwards and often seemed like his old self at other times. As she had not shared these disturbing events with her children, it was suggested that she do so in order to enlist more support and to apprise them of the true situation in case they needed to intervene on her behalf. When her husband's inexplicable behavior was explained in an understandable way, specifically pointing out that he had some capacity for good judgment but was prone to misinterpreting events due to his cognitive impairment, it was enlightening for her to have the pieces fit into place. It saddened her however, to think that she was losing her partner when it was pointed out that he might not be able any longer to appropriately digest decisions such as the dilemmas facing her daughter in divorce proceedings. He repeatedly became anxious and jumped to the conclusion that they would be financially drained after his daughter's divorce was final, and he appeared unable to empathize or see the situation from any perspective but his own irrationally fearful one. A concrete recommendation to stop discussing the details of the evolving divorce with him was made.

Mrs. Johnson's role in her marriage was also changing from being dependent upon the husband she had learned to rely upon as a full partner in making decisions and a sounding board for her own issues. Now, she needed to shift her view of him as someone at risk for misinterpretation of everyday

CASE VIGNETTE 11B (Continued)

circumstances with overly personalized, single-minded views of problems and, at worst, rage that bordered on violent outbursts. She agreed to be more selective about what she shared with him and to screen out the most complex dilemmas she faced and reserve those discussions for one of her children instead of Oscar. She now realized she needed to monitor her husband for escalating levels of frustration and to deftly monitor his medication supplies to verify that he was compliant with taking the antidepressant that not only helped his depression but minimized his potentially violent outbursts. Mrs. Johnson had just unknowingly crossed the line to becoming a caregiver.

Case Vignette 11: Discussion

Oscar showed symptoms that are indicators of ED despite having a relatively intact memory. Due to her frequent and intimate contact, Oscar's wife was in the best position to notice his "odd behavior." His children were prepared to dismiss their occasional observations of the same as "quirkiness" on their occasional visits. Regarding the sensitive family issue of her daughter's impending divorce and financial struggles, it is no surprise that Oscar's wife wanted to keep her husband's rantings quiet so as not to disturb the affected parties any more than necessary.

Mrs. Johnson tried to cope by hoping his behavior would blow over (denial) and realizing that she would rather sometimes be at work than come home (avoidance). Having Oscar's cognitive limitations explained to her was like having "all the cards placed on the table," she said. What had previously been inexplicable was now coming into focus as a pattern of behavior that had a root cause and for which there were better and worse ways to cope with. This new understanding was a great relief to her and the contingency plans that were discussed gave her new hope for being able to handle these new situations better. At the same time, however, to realize that she was losing (or had already lost) her intimate partnership with the man she married was causing her to "grieve the loss of the old role" (of being a full partner in a marriage). She was going through her own simultaneous role transition, where she realized she could no longer rely on him to be her sounding board to discuss her own thoughts about family matters such as her daughter's divorce. Although Oscar's wife was not the patient, the IPT-ci therapist encouraged her to share her new found information with her children for three reasons: (1) to have her begin to seek other sources of emotional support for the issues that were too complex for her husband to digest anymore, (2) to clue her children in about the possible future need to take action to protect her should his unchecked rage and disinhibition lead to violent behavior again, and (3) as his cognitive impairment could continue to progress in a worsening direction, it was time for the entire family to realize what was happening to be able to marshal all effective helpful resources. Her sense of needing to protect her husband's reputation in the eyes of his children was presented back to her as understandable and even noble but no longer practical, given the reality of his cognitive impairment as it was mapped out for her. This attempt to provide psychoeducation, clarification, and confrontation regarding the need to act for self-protection might seem as if the caregiver is being engaged in therapy as well; however, as was stated in the introduction, it can be argued that these steps were necessary for the benefit of the identified patient as well as for the caregiver's protection. The IPT-ci therapist thus straddles the interface and makes judgments about what is appropriate given the broadest view of the needs of both. In the case of the Johnsons, the action of excluding Oscar from the ongoing discussion of divorce maneuvering did have a noticeable benefit in reducing the number of times he became enraged and worried about money. This action was clearly adaptive as his input was not going to change the outcome of the divorce and repeatedly making him anxious by involving him in discussions that were too complex for him to process much less to contribute to in a meaningful way was merciful for him as well as helpful to his caregiving wife.

A subsequent vignette will illustrate another situation where the caregiver was referred for her own therapy. The option of obtaining her own therapy was suggested to Mrs. Johnson, but she politely declined stating that work exhausted her and that she did not want to commit to any new endeavors.

Oscar was treated with antidepressant medication of the selective serotonin reuptake inhibitors (SSRI) class of drugs, which has also been shown to decrease impulsivity and the intensity of his rage. Unfortunately, he had been deceptively noncompliant due to vague complaints of side effects and by his own statements that he thought the medication was too strong. He denied skipping medication doses even though his wife reported excess remaining pill counts to us. His medication regimen was thus changed to a longer-acting SSRI to try to provide a more consistent blood level even if he missed some doses. His wife agreed, in a separate individual meeting, to continue to check on his pill counts to try to verify whether he was being compliant, as he refused to let anyone dispense his medication for him. On subsequent sessions, she reported that he was being compliant and also noted a significant decline in the frequency of his rage attacks.

In Oscar's case, his neuropsychological testing suggested a pattern consistent with MCI at first, which progressed over two years of regular follow-up visits, such that retesting showed further deterioration that merited a diagnosis of dementia of the frontotemporal type. This diagnosis helped to explain the predominance of impulsivity and poor judgment that characterized his early clinical presentation as victims of frontotemporal dementia usually present with personality changes, impulsivity, and poor judgment early on more so than memory loss.

Oscar's wife seemed to need repeated explanations of his behavior indicative of ED (socially inappropriate comments, easily enraged, poor impulse control, limited problems-solving ability, and two instances of threatened violence toward her). Mrs. Johnson repeatedly told us that Oscar was very sensitive to being "bad mouthed" by his wife, and she implored us not to tell him the things she revealed for fear it would enrage him. In this case, safety issues seemed paramount and considerable time was taken in individual sessions with Mrs. Johnson to educate her about what was happening and to suggest not trying to win arguments with logical counter-arguments as Oscar's impaired insight could not grasp her attempts to do so. We recommend that she be very careful when he became enraged, and we tried to build in several contingencies for safety, namely: (1) stronger encouragement of compliance with medication, (2) we strongly advised having a revealing conversation with her children about the reality of the situation as we could envision that she might need their help in the near future. Up until this point, she had been shielding them from the reality to protect "their good image of their father," and (3) we rehearsed an escape strategy should he become threatening again that included making no attempt to argue, to remove herself to safety even it meant walking out of their home and then calling one of their children for help from a neighbors home. In that these behaviors were unpredictably rare and the fact that he would often profusely apologize afterwards and seem to forget that anything unusual happened hours later, she had been minimizing their seriousness and kept "hoping he would improve by coming for help." The degenerative nature of his illness needed to be repeatedly stressed with the likely expectation of more deterioration over time and a high likelihood that she would need more support and help to cope with this new reality.

It should be clarified that Mrs. Johnson did not immediately grasp all the psychoeducation presented to her and, although her own cognition appeared to be intact, she was not a quick learner such that the same issues needed to be reviewed repeatedly. When asked if one of her children might accompany them to a visit, she replied that they could not get off from work but could talk by phone. One son finally reached us by telephone, and the safety concerns were spelled out and his questions were answered. This son subsequently took it upon himself to stop by more frequently, ask his mother direct questions in private conversations and on one subsequent visit with them, he witnessed a "meltdown" after Oscar became fearful of bankruptcy after watching the news about the failing economy and subsequently began to blame his wife for frivolous spending. His son was instrumental in calming him back down after that incident. He is now a member of the support system.

We will return to the subject of the caregiver's own role transition in more detail in Chapter 10.

CASE VIGNETTE 12: AGING LEADER OF FAMILY-RUN BUSINESS WANES IN DECISION-MAKING POWER

As in the case of Oscar, the following vignette of Burt will illustrate another case of depression with a focus on role transition that initially improved with a combination of antidepressant medication and traditional IPT. As Burt's cognitive impairment continued to worsen, however, his mood worsened again, and this time it became clear that concerned family members needed to be involved.

Burt, an 82-year-old married male, was brought by his wife for treatment of depression. He owned his own structural materials company that he had begun from scratch 30 years earlier in which three of his sons now worked along with several other employees. He reported that he felt conflicted about retiring from the company as he saw his peers retire but he felt bored after half a day of playing golf with them and concluded that he would rather be working, which he enjoyed more.

He was engaged in traditional IPT around this role transition. Early in treatment, his wife asked for a private conversation during which she revealed her concern that he could no longer "hold his alcohol" and was driving aggressively at times such that his daughter was reluctant to have him pick up her kids from school any longer. Burt's IPT therapist insisted on bringing up the alcohol issue for discussion, and Burt stated that he mostly had a few drinks after playing golf with his peers, something he said he did not enjoy much as "they can sit for hours and talk about nothing." He was not aware that his driving was perceived as being unsafe by others but he vowed to stop drinking entirely. A subsequent check in with his wife who sometimes accompanied him to visits revealed that he was driving safely for several weeks and that he really did appear to stop drinking entirely.

Burt's IPT therapist helped him to weigh the pros and cons of continuing to work versus retire, and he seemed to favor continuing to work as it was not physically demanding and his longtime customers were like friends to him. The fabrication plant was also walking distance from his home such that even if he did decide to retire completely, he would always be in proximity to his business.

With continued clarification and some role playing around these themes, Burt made progress in depressive symptom resolution and agreed to long-term follow-up. He continued to remain alcohol free and his driving irregularity was no longer a concern for his wife. His neuropsychiatric testing showed MCI–amnestic type (cognitive impairment with predominant memory loss).

After 18 months of regular follow-up, Burt showed signs of a relapse in depressive symptoms. He agreed to meet with an IPT-ci therapist as the issue now seemed to be related to his memory difficulties and difficulty figuring out

CASE VIGNETTE 12 (Continued)

what to do with his business as he was aware he was making more errors and could no longer follow a project from start to finish due to failing memory. He kept saying that he just didn't know what to do but was vague about what the decision points were that he was struggling with. An offer to have a joint meeting with any concerned family was accepted and his wife and one son arrived for the meeting along with Burt.

When asked why they thought they were there for the family meeting, Burt said he felt ambivalent about "what to do with the business" as he was feeling more and more frustrated that his memory was failing him and that he often forgot important details that were costly mistakes. When asked if he thought his children were capable of running the business without him, he did not hesitate to say that they were technically able but he implied that he didn't think he was ready to trust their management skills.

His son confirmed that Burt was struggling with mistakes, and that they were double checking his work and went on to reveal that there was considerable tension between himself who acted as the accountant and his older brother who was the plant manager. He described his older brother as controlling and "needing to be reigned in" at which point Burt's wife added that she agreed and stated that this would not have been the case " in the old days" when her husband was more authoritative than he is now. Burt responded that he knew he "was not doing a good job" of being an authoritative leader because "he never believed in brow beating people." His son went on to say that he did not think his father was ready to walk away from the business and that he felt his father was not convinced that his children could run the business successfully without him. After further discussion, Burt's IPT-ci therapist pointed out that perhaps he was doing exactly what he needed to do—a slow and gradual withdrawal from the business. The meeting ended with the IPT-ci therapist asking Burt's son whether he felt it was wise to have a similar discussion between himself and his other siblings as everyone seemed to be in agreement that, due to his father's memory loss, he was having greater difficulty contributing in a direct way to the business but that there seemed to be more to it than that. His IPT-ci therapist also introduced the concept that there were other areas of cognitive impairment besides memory impairment reflected in his difficulty grasping the big picture, such as difficulty thinking of a way to bring the disagreeing parties (his children) to negotiate effectively (problem solving), and difficulty charting a definitive course for his graceful exit from the company he built (organizing and planning) even though he stated he wanted to become more completely retired. His son agreed that such a meeting was a good idea.

A follow-up individual session a week later with Burt alone revealed that he was relieved to have the issues "put on the table," and he admitted he had been avoiding the sibling-tension issues because he did not know how to handle them. He tended toward self-blame resulting in a palpable lowering of mood during the session. However, his IPT-ci therapist countered by pointing out that, in addition to his memory loss, his indecisiveness, organizational difficulties, and difficulty sifting through all the issues to seek the appropriate solution (difficulty grappling with complex situations) were part of the same changes in brain function that caused his memory loss and that these were not character flaws. It was pointed out that his wife recalled him to be much more authoritative in previous years and that these issues were complex and had no easy solutions, particularly in his current state of having a weakened memory and weakened problem-solving skills. This intervention seemed to provide some solace for him.

His IPT-ci therapist sought to buoy his flagging self-esteem by reminding him that it was the successful business he built that was providing a livelihood for all of his employees and three of his children.

A subsequent reconvening of the same four family members 3 months later (due to holidays and scheduling difficulties) revealed several things. His wife reported that Burt was much less irritable on his new antidepressant and that he was going into the plant less often and trying to take up a hobby (ceramics). She said she would prefer that he strive for a 50–50 split of his time, however. His son revealed that the siblings had indeed talked and that they were assuming more of the management burden so their father would not have to feel obligated to do so but that he was welcome to come and go as he pleased while having little or no true responsibility. On the other hand, Burt's son defended his father's attempts to participate (with help) in the business as useful and fruitful even when Burt down played his own efforts as barely significant. Burt's son also reported that the siblings were working together better and that "they all had to accept that they had different personalities."

Case Vignette 12: Discussion

The extent to which Burt's executive dysfunction was interfering with his work abilities on multiple levels did not become clear until the first family meeting was held. At that meeting, the impact of his memory loss on job performance was clearly evident (by his own admission) even though his children were loosely checking his work. The larger issue was the degree to which he was unable to take charge any longer, sit his employees down to negotiate disputes, and more clearly define job descriptions and who answered to whom, and so on. His eldest son was behaving like a self-appointed *de-facto* CEO, and even though he may be the best qualified for that position, it was never officially voted upon or agreed to which rankled at least Burt's younger son in attendance at the meeting. Burt, his children, and his wife all wanted the same thing: a graceful exit for Burt that allowed him to preserve some role as a figurehead, but no one seemed to know how to make it happen. They all wanted Burt to be able to retire with dignity, be able to come and go as he pleased in the business, and feel secure that the business he created would continue to thrive and that his children running it would do so amicably.

His IPT-ci therapist could quickly see that Burt was not going to be able to plan and execute an exit strategy by himself and that he tended to blame himself and suffered lowered self-esteem because he was not sure what to do. He knew things were not going well in the business and between his children in the company. His accompanying son may have come because he had an axe to grind, but he was also clearly worried about his father and also the breakdown in communication within the business. His IPT-ci therapist sought to clarify the steps required to allow Burt to bow out gracefully over time (his stated wish) while maintaining his self-esteem and beyond that, to help all parties celebrate the fact that he had built such a successful business that provided a livelihood for three of his children as well as several other employees. Trying to speak plainly for all to hear, his IPT-ci therapist pointed out that, just as Burt's memory had faded (which he was the first to admit), his decision-making capacity had similarly suffered from the same degenerative process (probably Alzheimer's disease), and it was now time for his children to step up and figure out how to work together to run the business so that their father could slowly pull back further and retire completely when he felt ready to. The follow-up family meeting showed that things were beginning to happen. Better communication was taking place between his children and it was clear that the entire family agreed to pull together for Burt's sake if not for that of their own sakes if the business was to remain a successful one.

A monthly one-to-one maintenance IPT-ci session with Burt revealed that he was relieved that the IPT-ci therapist had expressed what he could not figure out how to say. Furthermore, these sessions provided an opportunity to challenge his readiness to blame himself for failing to take action sooner by providing psychoeducation about the deleterious effects of early dementia on multiple abilities besides memory that were beyond his control. Burt seemed relieved after discussions along these lines and together with the effect of his antidepressant medication, his mood remained in the normal range.

A subsequent follow-up meeting with Burt alone, a year from the first family meeting, revealed that now he only visited the business for 30 min daily and his "job" was to bring in the mail and take checks to the bank for deposit. Although Burt was now taking cholinergic-enhancer medication, he complained that his memory and his word-finding ability were getting worse. He also wanted to talk about side effects of his medication, which was subsequently adjusted. He seemed to appreciate the nonjudgmental aspect of the meeting and reported that his mood had remained stable despite worsening cognitive ability. He also said he was now working on a hobby (ceramics) that he never took time for previously. Burt and his IPT-ci therapist agreed to keep meeting periodically to review his progress and he vowed to bring in an example of the pottery he created.

More details of specific IPT-ci interventions will be presented with additional case vignettes in Section II.

The Additive Disinhibiting Role of Alcohol

The case of Burt outlined earlier demonstrated the added negative effect of alcohol on driving safety, which Burt was not even aware of. Alcohol use and other drugs (prescribed or illicit) can worsen ED by exaggerating the aforementioned symptoms. Alcohol or sedative drugs can trigger problem behaviors by anesthetizing areas of the brain that normally serve to inhibit impulses.

A good illustrative example of how normal inhibition develops is to observe a child who is scolded by his parent at age two for reaching out and grabbing the candy from the child next to him. He or she is being taught "civilized" or "socially expected" behavior. Much of being "civilized" is inhibiting impulse gratification of one desire or other (food, attention, novelty, sexual gratification, etc.). Developing this inhibition is a cognitive learning process that is largely "housed" in the frontal lobes of the brain. Anatomically, the frontal lobes of the brain are the largest part of the human brain that distinguishes humans from the rest of the animal kingdom. The frontal lobes carry out the highest order intellectual functions such as planning, forethought, insight, complex problem solving, multitasking, and the discernment of nuances of difference (judgment). Damage to the circuits in the frontal lobes can bring disinhibition usually characterized as socially inappropriate behaviors or language (swearing inappropriately, tactless comments and sometimes physical overreactions, rage or even violent acts as in the case of Oscar). In a vulnerable individual, alcohol can further anesthetize the inhibitory capacity of the frontal lobe and exacerbate socially inappropriate behavior.

Difficulty of Recognizing Gradual Onset

As the onset of ED is usually gradual and often sporadic in its expression, it is often seen by family and friends as an exaggeration of someone's personality quirks or it is attributed to the aging process. Family, friends, and coworkers often do not see the changes as evidence of an underlying brain disease (and why should they if they have never been educated about the phenomenon). More often, these symptoms are taken at face value by others who have difficulty understanding why the person with ED "won't see" the problem being pointed out to them. Family members try making logical arguments for changing perceptions or behaviors, which are not well-received by persons with ED. Family members then erroneously conclude that a spouse, for example, is falling out of love with them or has "turned mean" inexplicably. Circumstances that produce increased stress—such as a requirement to multitask or constantly shift attention or when the individual is distracted by a noisy environment or feels pressured or hurried—can also exacerbate the symptoms of ED.

Executive Dysfunction and Driving Safety

Driving, of course, can be impaired by a variety of other co-occurring factors in addition to cognitive impairment such as failing eyesight or hearing, slowed reflexes, and limited movement in joints due to painful arthritis. For these reasons, doctors are mandated by law to report patients to the state licensing authorities who they believe might pose a risk for operating a motor vehicle unsafely. Such a report mandates that the afflicted person be required to present themselves for a formal driving reevaluation. If the person is deemed unsafe at such a formal retest, then the privilege granted by each state to drive is revoked for the greater welfare of society. If an afflicted person agrees to stop driving (verified by family members), then he or she can be spared the embarrassment of being reported. Sometimes, however, the afflicted person is eager to be retested "to prove" that he or she is a good driver. If the formal testing results in a loss of licensure, this displaces the blame from the family or the treating clinician to the state agency. Further suggestions for approaching the subject of driving are listed in Table 9.2 in Chapter 9.

The Value of Psychoeducation in Executive Dysfunction

When family members are adequately educated about ED and provided with practical techniques such as those incorporated in IPT-ci, they become more tolerant and understanding and show greater patience and less hostility and resentment as they learn to avoid fruitless confrontations. When the true reality of the cognitive impairment is mapped out for both the victim and the caregivers in understandable terms, they may become partners for bringing about necessary changes. They can limit decompensation by seeking greater clarity or simplicity and by choosing tasks that the identified patient can reasonably be expected to handle. When safety is a concern such as in driving and the individual exhibits recurring lapses in judgment, sometimes brainstorming about possible solutions is required to limit access to potential dangers as the following brief vignette will illustrate.

Another feature of ED is a decline in problem-solving ability. This can be inferred in a variety of ways, for example, by increased difficulty balancing check books, understanding instructions, or operating appliances. Driving ability can be similarly impaired if too many decisions are required at once. In situations like driving, such distractions or pressure can create overreaction, for example, as in "road rage" that can lead to accidents as the afflicted person is less able to inhibit inappropriate action or couple the right amount of action/restraint with good judgment. Someone with ED may have had a good driving record and may seem like a fine driver going to routine places but may decompensate, panic, become confused, or overreact when driving in unfamiliar surroundings or if an unusual situation arises that demands quick and decisive action such as avoiding a child running after a ball in the street without swerving into oncoming traffic.

CASE VIGNETTE 13: SAFETY HAZARD DUE TO EXECUTIVE DYSFUNCTION

George, a 74-year-old retired salesman, was brought by his wife for treatment of depression and to make sure his "medications for depression were correct." George had endured two heart attacks and had been diagnosed with narrowed carotid arteries as well, one of which had been surgically reopened to try to reestablish good blood flow to the brain. George clearly had bad arteriovascular disease involving his coronary and cerebral vessels. He showed two small strokes on his MRI scan and more white matter hyperintensities than would be expected for a man his age. These findings along with his patchy memory loss, change in personality, frequent bouts of giddiness, and lack of insight or concern for the same worries his wife articulated, all pointed to a diagnosis of early vascular dementia. He spent his day "puttering" around the yard raking leaves or working in his garden or his basement workshop as he had done ever since he retired from a job as a plant foreman. Occasionally George would have a "temper outburst" according to his wife but mostly he was affable, pleasant, and very willing to comply with the agenda his wife set.

One day, however, the power went off in their home while his wife was reading a magazine and she thought that a fuse had blown. She was shocked and frightened to learn instead that her husband had decided to do a rewiring project in the basement and had cut through several live electrical wires with handheld wire cutters that showered sparks, doused the lights upstairs, and fortunately, did not electrocute him. When confronted with the safety concern for his actions, George reported with an unfazed demeanor that "we always did it that way at the plant." With the realization that George's judgment was impaired and that this carried serious safety risks for him, a decision by his family to provide more supervision for Georges' "puttering" was discussed at length.

On another occasion, George's wife reported an incident that troubled her greatly when George was caught "shoplifting" in their local hardware store. It seems that a clerk saw him put a small electrical part in his pocket and confronted him. George stated that he just "parked it" there temporarily while he continued to browse. He had the money to pay for the part in his wallet and since its cost was less than a dollar, he was let go with a warning. His wife was mortified by the shame of the near miss of a charge of shoplifting and brought him back to the psychiatrist to "do something." George was unfazed again and said he did not even recall putting the object

CASE VIGNETTE 13 (Continued)

into his pocket. He further denied intending to walk out without paying, and he could not understand why everyone seemed to be so upset about such an object of little value. George was obviously missing the "big picture" that shoplifting is a prosecutable act regardless of the value of the item being pilfered and that claiming to be unaware it was in one's pocket (even if a true statement) was not a defensible position in the eyes of the law. Furthermore, as the local hardware store was within walking distance from his home, it was likely that they would be scrutinizing him carefully in future visits and take action for a second offense.

The possibility of accompanying George on shopping trips was discussed at length but he lived within walking distance of the hardware store and his wife's painful arthritis precluded her from always going with him. A pattern was emerging of George unwittingly and automatically putting several kinds of objects he encountered into his pockets as his wife often found an odd assortment there when she did the laundry. A month later, she was again upset when she found that he had again carried a new object home from the hardware store, this time without getting caught. Again we brainstormed about more supervision or restrictions on his travels alone but a simple solution was tried, which proved to be effective. George's wife agreed to try sewing George's pants pockets shut. George did not complain about the inconvenience and no more objects were inadvertently shoplifted.

Case Vignette 13: Discussion

These two examples of ED with poor insight into the consequences of their actions (Oscar and George) illustrate the safety issues that sometimes demand immediate protective action. They also demonstrate the risk of having inappropriate behavior create consequences in the real world, where other parties are neither cognizant of the underlying problem nor likely to be sympathetic. If George were prosecuted for shoplifting, it is likely that a letter from his physician would vindicate him but the length of the proceedings would have been highly stressful and embarrassing for George's wife.

The manner in which these behavior patterns were handled by the IPT-ci therapist will be broken down into steps in greater detail in Section II. It is noteworthy that all three case vignettes of Oscar, Burt, and George presented with depression. The case vignettes of Oscar and Burt showed that the utilization of traditional IPT was helpful to engage and work with them individually at first, but with continued progression of cognitive impairment, the additional IPT modifications of providing psychoeducation about ED for their family members was crucial for helping all parties to cope effectively with further deterioration and problem behaviors.

Neither Oscar nor George retained much insight for being able to utilize individual work in IPT-ci although a good working rapport with the therapist was maintained. Their visits to the IPT-ci therapist seemed to be a routine pleasantry for them during which the bulk of the intervention was directed at providing psychoeducation to their family members and brainstorming with them to solve problems. The case of Burt, however, was more focused on his own view of the challenge of figuring out how to retire gracefully and feel that the company he built was in good hands with his children. Despite his flagging memory and evidence of ED, Burt still possessed enough insight to digest the points being presented to him, and he appreciated the clarifying help with solving his problem (what to do about his business).

Section II

IPT for Cognitive Impairment (IPT-ci)

7
IPT-ci Basics

Now that we have reviewed the basic principles of IPT for use with cognitively intact elders as well as provided a basic review of gerontologic/geriatric medicine, depression, cognitive impairment, and executive dysfunction, we will now turn our attention to specific techniques for using IPT for cognitively impaired patients. IPT-ci is intended for delivery within primary care settings, multi-disciplinary geriatric centers, or by solo practitioners.

Conceptualization of IPT-ci Treatment

IPT-ci treatment can be conceptualized in three phases:

1. The first phase consists of the initial evaluation, the assessment of depression and cognitive status in the identified patient, the work-up for underlying biological causes of cognitive decline and/or depression, trials of any psychotropic medications, ample psychoeducation for both the identified patient and any concerned family/caregivers (this can be done jointly or separately as clinical judgment dictates), completing the interpersonal inventory, giving the patient the "sick role," and deciding upon and articulating a focus or problem area for the treatment intervention.
2. In the second phase, the identified problem area(s) is addressed as in traditional IPT, with the addition of separate sessions and possibly joint sessions for caregivers at the IPT-ci therapist's discretion for continuous assessment of their own coping and to provide a joint forum for role dispute resolution. The ultimate goal is to achieve a steady state where cognitive impairment and depression in the identified patient are maximally treated and where both patient and caregivers are fully educated and have found optimized coping strategies that seem to work best for them (this may be an evolving process). Specific techniques for coping with role transitions, role disputes, interpersonal deficits/sensitivity, or unresolved grief are utilized here. Achieving the goal of depressive symptom resolution through traditional IPT techniques remains a goal in this phase if the identified patient is cognitively able to participate.

3. In contrast to the third phase in traditional IPT where the therapist works toward consolidating gains and preparing the patient for termination (except when there is a plan for maintenance IPT in cases of recurring depressions), the third phase of IPT-ci is a long-term follow-up or maintenance phase where "check-up" sessions are regularly scheduled using the IPT-ci therapist's judgment to gauge sustainability of the treatment strategy, to maintain surveillance for a recurrence of depression, and to be alert for new problems or further declines in cognitive status that require a reevaluation or new strategy to maximize coping commensurate with the increased impairment. It is also within this longer-term follow-up phase that the IPT-ci therapist gradually introduces such topics as end-of-life issues, advanced directives, consideration of long-term care settings when indicated, and other preparations for the final phase of life.

Table 7.1 summarizes the 17 basic steps of IPT-ci that will serve as a template for the details and illustrative case vignettes to follow. The reader is encouraged to refer back to this table as often as necessary to remain oriented to the overarching strategy for delivering IPT-ci to patients with cognitive impairment who may or may not have comorbid depression. Concerned family members or caregivers need to be assessed in their own right for suitability to be effective caregivers, as well as to incorporate them directly into the treatment plan for a given patient. Throughout treatment, the therapist should monitor caregivers' stress levels, which can directly impact their ability to provide effective care.

Table 7.1 A Summary of Steps Involved in IPT-ci

1. **Engage the identified patient** in a comfortable setting with a conversational style that allows you to asses his or her understanding of why he or she is there and his or her view of his or her problem(s). Respectfully engage the patient and solicit his or her story directly by asking any accompanying family or caregivers to allow you to obtain their opinions after you have first engaged the identified patient.
2. **Observe family or caregivers and the identified patient** together either in the waiting room or in the interviewing office. Inform both the identified patient and his or her family that the evaluation and the treatment process will take place in steps. Set limited goals for any one visit, particularly the initial one. The goals for the first meeting are to establish rapport with the identified patient and caregivers and agree to begin a work-up to search for causes of cognitive impairment or depression (such as screening blood work for thyroid dysfunction or vitamin B12 deficiency). You may perhaps want to gather old records

Table 7.1 (Continued)

or speak with other clinicians or concerned family members (with permission) for additional background information. Standardized screening instruments to assess depression severity or cognitive status can serve as a benchmark with which to gauge progress or further decline.

3. **Model respectful behavior** toward the identified patient for caregivers or concerned family members to observe. Consider interviewing accompanying family or caregivers separately so they can speak freely. Use active listening to allow the patient and caregiver to paint a picture of the environment that is the stage where the problem behaviors, concerns, or depressive symptoms are played out in order to gather information useful for developing a treatment strategy. Define your role for the identified patient and the family as an educator and a facilitator whose goal is to help both the identified patient and his or her family/caregivers to understand the nature of the problems being presented, to search for underlying treatable causes, and to collaborate with primary care physicians or psychiatrists as necessary to help monitor the effects of any prescribed psychotropic medication. Overall goals are to use principles of traditional IPT or IPT-ci modifications to minimize depressive symptoms, maintain the dignity and self-esteem of the identified patient (particularly in the face of declining cognitive ability), and to assist the problem-solving process for both patient and caregiver by acting as a bridge over any gulf of misunderstanding (this may require individual or joint role dispute resolution sessions).

4. **Complete the interpersonal inventory** to include anyone the patient lives with, any other support system members (such as church members, friends, coworkers, or neighbors), and all family members, particularly those who are engaged in treatment decisions (whether or not they live close by). As in traditional IPT, ask questions that provide information about the quality of these relationships and take note of any nonreciprocal role expectations on the part of either the identified patient or any caregivers.

5. **Begin psychoeducation from the first encounter**. Provide ample psychoeducation to both the identified patient and his or her family (perhaps in separate sessions as indicated) about both depression and/or cognitive impairment/dementia in terms they can understand and in digestible units. Use reading material or videotapes to enhance their knowledge as appropriate.

6. **Complete a thorough psychiatric interview**. Thoroughly evaluate depressive symptoms and cognitive impairment symptoms and their context. Obtain the patient's prior history, family history, social history,

Table 7.1 (Continued)

and substance abuse history. Conduct a mental status examination. Listen for exacerbating circumstances such as worsening problem behaviors in the early evening (sundowning), when tired or stressed, when in noisy or overly busy environments, when required to multitask, or when under the influence of alcohol or drugs or even prescribed medications that may produce a disinhibiting effect such as benzodiazepines, for example, alprazolam (Xanax®) or clonazepam (Klonopin®).

7. **Consider a thorough medical evaluation** to rule out occult medical problems such as commonly encountered hypothyroidism and low vitamin B12 levels and less commonly, undiagnosed cancer, small strokes, normal pressure hydrocephalus (NPH), or early Parkinson's disease. Collaborate with medical specialists as needed.
8. **Assess cognitive status in the context of the observed behaviors** to consider embarking on a dementia work-up to look for treatable causes of cognitive impairment with referrals to neurology, brain imaging, or other specialized evaluation as indicated.
9. **Consider neuropsychological testing**. Use objective measures such as rating scales and standardized testing to gather reliable assessments and take time to patiently explain the results in plain language using examples or metaphors to help all concerned parties to understand the interpretation of the results. An example of a useful metaphor for a retired mechanic with signs of psychomotor retardation might be an analogy in which his or her brain seems to be functioning as poorly as a car engine running on only four out of six cylinders.
10. **Be alert for signs of executive dysfunction** (as outlined in detail in Chapter 6), particularly safety issues that might require immediate action.
11. **Consider the need for psychotropic medication** for depression, cognitive enhancement, anxiety, insomnia, agitation, impulsivity, or psychosis. Refer the patient to the appropriate physician (preferably a psychiatrist with geriatric subspecialization, if available) to obtain such medications and have them monitored with regular follow-up visits.
12. **Acknowledge that concerned family or friends may be going through their own role transition** as they begin to worry and provide more surveillance or assistance to the identified patient when they realize that the identified patient is showing signs of worsening cognitive impairment. Be alert for signs that caregivers may be strained or that their caregiving ability may be compromised. Assess caregivers' level of capability to give care that might be impacted by their own state of mental and physical health, level of education, or familiarity with the common problems in the identified patient's age

Table 7.1 (Continued)

group (even healthcare workers do not necessarily have experience in this area). Be a resource for appropriate referrals for identified patients who need specialty care or for caregivers who may need their own treatment for medical, psychiatric, or substance abuse problems or legal advice.

13. **Keep adequate notes of salient issues such that patients with memory impairment can be reminded of therapeutic work done in previous sessions** (as well as any caregivers who were involved in the relevant issues). Consider providing written notes to assist the patient and/or caregiver in trying to carry out suggestions for new strategies between treatment sessions. Formulate a problem area that you plan to focus on such as the "role transitions" precipitated by cognitive impairment-induced losses in social functioning or "role disputes" that can arise, for example, from caregivers attributing executive impairment to willful opposition. Explain to both the identified patient and the caregiver that you are assigning the "sick role" to the identified patient as he or she is currently unable to carry out all of his or her prior activities or duties even though he or she may wish that he or she could. Explain to all that when any accompanying depression is adequately treated, the identified patient's cognitive functioning may improve although it may not return to its lifetime peak.

14. **Listen to caregiver concerns and try to help them understand what is happening with the identified patient.** (Be clear that your allegiance is to the identified patient.) Be respectful of the autonomy of the identified patient and the caregiver(s). Explore optimal ways of coping (brainstorming) in separate sessions with the patient and caregiver(s) if indicated. If it is determined that a caregiver needs his or her own psychotherapy, mental health, or substance abuse treatment, then assist him or her to obtain it. Advocate for the patient's best interests in the event there is a suspicion of caregiver abuse, which in rare circumstances, might require taking protective action against the caregiver.

15. **Explain the structure of IPT-ci**, which involves a flexible combination of individual or joint meetings with concerned family members or caregivers to allow you (the IPT-ci therapist) to obtain all necessary information to help all involved to work together to reach a steady state of maximized coping for both the identified patient and those who are concerned about the patient. The IPT-ci therapist uses clinical judgment and flexibility in deciding the frequency of follow-up visits by taking into account the intensity of problem behaviors or depression severity as well as the need for logistical and

Table 7.1 (Continued)

practical problem solving. Ideally, regular sessions will take place weekly to maintain therapeutic momentum initially. Interim phone contacts can further maintain momentum when weekly visits are not possible. Staying in contact with caregivers separately by phone or by email can enhance their coping ability and provide regular updates about their observations on the identified patient's status.

16. **Work to establish a "steady state."** A steady state is reached when an adequate work-up is completed for possible causes of cognitive impairment, when any appropriate cognitive-enhancing drug therapy is optimized, when depressive symptoms are minimized, when both the identified patient and caregivers have been thoroughly educated, and when any role transitions, role disputes, or bereavement issues have been worked through. The next phase of treatment would be similar to the maintenance phase of IPT used for patients with chronic or recurrent depression at high risk for relapse. In IPT-ci, however, there is also the added element of possible further cognitive decline over time that may produce new behavioral symptoms, a resurgence of depressive symptoms, or new role disputes. A flexible schedule of follow-up needs to be maintained to allow the IPT-ci therapist to be able to reassess the identified patient's status and the adequacy of the caregiver's ability to adapt. The steady state phase may be marked by months or years of relative stability or minimal change during which follow-up visit frequency might extend to three or four month intervals for the remaining lifetime of the identified patient. If a new crisis emerges or if decision points seem to be approaching such as the need to stop driving or move to an assisted living facility, more frequent visits should be resumed until a new steady state is achieved.

17. **Prepare all parties for future contingencies**. During this steady state phase, the IPT-ci therapist should look for an opportunity to diplomatically prepare all capable of participating in a discussion of possible anticipated needs for future contingencies if cognitive impairment progresses on a downhill course. This might involve a move to a more supportive setting, creating power of attorney agreements, or the institution of safety measures such as using medication organizers, ceasing to drive or use dangerous appliances, or hiring part-time home health aides or drivers. Every person should be offered advance directives and legal counsel may be required to document the identified patient's wishes accurately as well as to ensure that last will and testaments are in order.

Strategies for Caregivers

Focusing on the caregivers directly, Table 7.2 outlines eight pragmatic strategies for improved coping.

Table 7.2 Basic Strategies for Caregivers

1. Keep it simple. Implement new strategies in steps.
2. Search for strategies or activities that enhance the self-esteem in the identified patient.
3. Consider altering former favorite activities of the identified patient so he or she can continue participation to some degree. For example, if a complicated game of cards was enjoyed but is now too taxing, try switching to a simpler game, or if socializing with friends has fallen off, consider arranging simpler meetings such as a one-on-one lunches in a quiet setting where there is less distraction and schedule such meetings when the identified patient is better rested, earlier in the day. Encourage substitute activities when cognitive impairment precludes the old favorites. Keep in mind that as cognitive impairment progresses, foresight and planning, as well as memory, can be diminished such that activity levels often decline unless an external organizing force encourages activity. The concept of caregiver as "coach" may be very helpful in these circumstances.
4. Praise the identified patient for attempting tasks and for all successes, however small (such as helping with household tasks, cooking, or any crafts or artwork he or she might produce).
5. De-emphasize the negatives even though you acknowledge them. (Those afflicted with cognitive impairment often feel less useful or like they are failing and as such, enhancing self-esteem often requires an active stance to counteract this tendency. Conversely, be aware that critical remarks can be very damaging to self-esteem.)
6. Offer choices whenever possible but avoid the appearance of micromanaging the identified patient's life as that feels stifling and infantilizing to all except those with dependent personality traits. For example, simply ask the patient to choose between two options for food choices, timings of events, or preferences for recreational activities. Also, ask the patient's opinion on topics he or she can still grasp or follow.
7. Thank the patient for his or her help for any contribution to everyday activities. Keep reminding the identified patient of his or her (important) role in the family and his or her valued input.

Table 7.2 (Continued)

8. Structure ample time for yourself (the caregiver) to "recharge your own batteries." Continue activities you enjoy even if the identified patient can no longer participate. Learn effective strategies for taking care of yourself simultaneously. These efforts will also indirectly benefit the identified patient. Establishing "coverage" by other parties (family or paid help) to allow regular time for caregiver's autonomous activities may need to be negotiated.

The Initial Interview

One cannot assume anything about the functioning of a patient based solely on his or her age. If patients come alone or if they insist on being seen alone despite accompanying family, the IPT-ci therapist accedes to their wish, out of respect, at the first meeting. Although the bulk of this manual is focused upon modifications of IPT to better handle the complexities of depression coexisting with cognitive impairment in the identified patients along with the reactions of their concerned family members, for those elders who are fully intact cognitively and are presenting for help with depression alone, traditional IPT may be delivered on a one-to-one basis as it was originally conceptualized. The modifications of IPT-ci built upon the tenets of traditional IPT can be subsequently "rolled out" to accommodate the special needs of the patient who may subsequently show signs of cognitive impairment with or without depression.

Involving the family in the therapeutic process in IPT-ci may be crucial when treating a patient with cognitive impairment. In cases of "pure depression" (with no cognitive impairment evident), the active involvement of family members may still be helpful to educate them about the effects of depression and to consider how their efforts might help the identified patient make progress to recover from depression more fully and expediently. It is important to note that family members of depressed relatives are often just as stressed as those who live with a cognitively impaired relative and may need education, empathy, and perhaps even their own referral for treatment if they themselves are depressed from the stress of living with a severely depressed person. As per traditional IPT, role disputes may also be contributing to the depression in the identified patient, which may require further exploration.

In the initial interview with a new geriatric patient, it is important to set up some basic "rules of engagement," particularly when family members accompany the patient. It is a good idea to always greet the patient first and introduce yourself. This is important for two reasons: It models a respectful relationship for the family to observe between the clinician and the identified patient, and particularly for patients who are already feeling insecure or are chafing at the attempts by caregivers to exercise greater control over their lives. It also sends a message to the identified patients that you take them and what they have to say as seriously as those who may have brought them for help.

On a pragmatic level, assessing hearing acuity is important for good communication. Position your chair on the side of the "best ear" so you can be sure the patient is hearing you clearly. Hearing loss is a frequent problem among older individuals and when combined with cognitive impairment, it can easily lead to misinterpretation and even paranoia. Inexpensive hearing amplifiers can be purchased for use in guaranteeing effective communication in office settings. (They consist of head phones tethered to a handheld microphone with a battery-powered amplifier.)

Clearly identifying yourself and what you intend to do is important. After making patients comfortable with small talk, the IPT-ci therapist usually begins by asking where they are from. Their answer gives the therapist a quick assessment of the smoothness with which they handle the details of explaining the specifics of their particular geographic locale. The therapist may also ask how long it took them to travel to the therapist's office and gauge their sense of time as well. If they appear confused or make an incorrect statement, it is always interesting to note whether they self-correct, defer to family members, or whether family members jump in to answer for the patient as if they may have done so habitually and have concluded on their own that the therapist needs to have accurate information to proceed. After noting whether family members take the initiative to answer, the therapist may quickly put up an open hand and gently silence them, simply stating that the therapist would like the patient to answer for now but that the therapist wants to hear from them too a little later. Usually, family members quickly catch on.

After the comfort level has increased with the opening orientation questions (some ice-breaking humor doesn't hurt either), the IPT-ci therapist asks patients what their understanding is of why they are here today. Again, if they are not sure or look to family members to explain, the therapist makes a mental note but gently presses them to tell why *they* think they are there. If they mention their memory being a problem, the therapist asks them to elaborate and give their opinion of the problem or problems. If they say their family brought them, the therapist responds that he or she will be asking their family members for their input later but would like to hear directly from them first.

If depression is mentioned as patients' reason for coming, the therapist launches into a more thorough history of depression over their lifetime, whether they recognize any particular stressors that may have precipitated or contributed to the depression, whether medications have helped this episode or those in the past, and so on. Assessing for other risk factors such as substance abuse or comorbid anxiety disorders is also warranted at this time to gain the most complete picture possible. If understanding depression is unfamiliar territory, then this may be the time to digress in explaining that major depression is an illness that affects the functioning of the body as well as the mind. Check for physical manifestations of depression, such as symptoms of sleep and appetite disturbance and exacerbation of pain symptoms. Also conduct a thorough exploration of any suicidal thoughts or behaviors (as outlined in Chapters 4). Obtaining this information is certainly part of a complete psychiatric history and doing so in a manner consistent with the IPT-ci approach can go a long way to setting the stage for good rapport, trust, and cooperation between the identified patient, caregivers, and the IPT-ci therapist.

After the IPT-ci therapist obtains a good idea of the identified patients' impression of their reason for coming, the therapist asks their permission to speak to their accompanying family members to help the therapist understand what they think the problem(s) are. If the patient hesitates or starts to say something like "well, they wanted me to come because they think I should stop driving," then the therapist makes a mental note of this potential role dispute and continues to allow the patient to elaborate on his or her view of the situation at length before allowing any family to outline their views (which the therapist suspects will differ). While the patient is talking, the therapist also makes mental notes of the facial expressions and gestures of the accompanying family members. (The author has seen family members grimace, roll their eyes, or minimally shake their heads as if to say in effect "that's not true.") If family members try to interrupt to correct the patient, the therapist again gently reassures them that he or she certainly wants to hear from them too, in a few minutes.

If patients show embarrassment or appear to feel badly about the cognitive deficits that have been noted by themselves or others, the therapist may intervene with an educational approach that places, for example, memory loss in perspective as a problem faced by many in the same age range and for which a medical cause should be sought. Some patients are already terrified of the term "Alzheimer's disease" as it has connotations of losing independence, dignity, and being placed in a nursing home. Without sugar-coating the reality of Alzheimer's disease, it is often reassuring to point out to patients that their diagnosis is not certain as yet, and that even if a diagnosis of Alzheimer's is the result of the investigation, an optimistic tone can counter the tendency to catastrophize by pointing out that the progression of Alzheimer's disease is usually slow, that medications (like Aricept®) may slow the progression further and that the illness sometimes changes very little from year to year such that many people who are diagnosed with Alzheimer's disease continue many of their activities more or less as they have been doing them for many years to come.

Further details that constitute a good initial psychiatric evaluation should be obtained including pertinent past psychiatric history, social history, family history, substance abuse history, pertinent medical history (if not already available), and a complete mental status examination. Administering the Mini–Mental Status Examination (MMSE) is standard practice and may be augmented with other neuropsychiatric assessments as appropriate. Recently, we have been using the Montreal Cognitive Assessment (MoCA) described in detail in Chapter 6 as it is more sensitive to executive dysfunction. Be sensitive to the patient's reaction to feelings of embarrassment for not knowing an answer, particularly when tested with family members present in the room. If you sense the identified patient is embarrassed by being assessed, you might consider finessing the situation by announcing that your standard operating procedure at this point is to ask the family to wait in the waiting room while you administer some brief memory testing and then you will then ask them to rejoin. Sometimes, patients are not bothered by the testing and would prefer that their family remain as they are tested. Patients who make errors in the testing may minimize them or laugh them off, while family members sometimes express shock or dismay when confronted with the fact that their loved one is unable to answer correctly. When a strong reaction from family members is witnessed, it may be necessary to see them separately to draw them out and hear about their worst fears, particularly their worries for the future. Personality traits obviously play a role in the demonstrable coping skills of the caregivers in the face of depression and/or declining cognitive capacity in their loved one as the following vignette will illustrate.

CASE VIGNETTE 14: ACCOMPANYING DAUGHTER BECAME DISTRAUGHT

During an evaluation of a 78-year-old widowed woman, her accompanying daughter (an only child) burst into tears when it was revealed that her mother's pattern of memory impairment seemed to fit the usual pattern of Alzheimer's disease. She sobbed with deep heaves for 15 min during which time she revealed that she had suspected as much and had unsuccessfully tried to remediate her mother's obvious deficits by having her read from "Dick and Jane" first-grade reading books as well as coaching her all the way to their appointment in the car so that she might retain the correct date long enough to "pass" the test. The identified patient was not upset by the testing but her daughter required a great deal of psychoeducation, patience, and empathy to assist her to be able to digest the news and to begin to make reasonable plans for coping with it.

If a dementia work-up is indicated by either preliminary tests or by other history (such as getting lost in familiar places or forgetting names of grandchildren), making arrangements for doing so are recommended to the patient and the family. The IPT-ci therapist promises to help everyone understand what the implications of the test results are and to "brainstorm" about the best options to proceed to cope with the identified problems. This process is outlined as a learning process for all involved: the patient, the family, and also the therapist, who wishes to learn more to be as helpful as possible. Making arrangements for further testing (screening blood test or neuropsychological testing) may be a reasonable place to conclude a first encounter.

Psychoeducation Begins Early in the Evaluation Process

Psychoeducation about depression and cognitive impairment/dementia begins with the first contact as illustrated earlier. Patients and caregivers bring a variety of backgrounds and education levels, therefore, tailoring an educational approach to a particular patient or group requires some judgment about which issues are hottest and must be dealt with immediately to defuse an immediately discernable role dispute or misinformation that needs to be corrected upfront as it could potentially lead to inappropriate reactions. Obviously, a limited amount of education can be digested at one setting, and judgment is also required for devising a strategy in your own mind for how great the educational need is and a game plan for delivering it in the manner most appropriate. A statement conveying some reassurance that you have worked with other people with similar problems and that there are many strategies you are happy to teach them over time is a good idea, followed by some language such as "but for today, lets concentrate on getting started by..." and then proceeding with a concrete plan, for example, to obtain the records of previous treatment, obtain blood tests as part of a dementia work-up, or begin an antidepressant if indicated by a qualified healthcare practitioner. It is a good idea to point out that the learning process will unfold over time and that you are optimistic that greater understanding will help all parties involved to cope better, and hopefully to experience improved mood and a recovery of some lost functioning due to either the depression or a sense of being demoralized by losing cognitive ability or both.

Common "burning questions" from either patients or family members should be anticipated such as "Can you tell us if this is Alzheimer's disease?" "How long will it take to see any improvement?" "How fast are they going to go downhill?" or "Does AD run in families?" Hopefully, by this point, the psychoeducation you have delivered in "lesson one" has established you in the eyes of the group as a knowledgeable and patient advocate such that you can either answer their question in a concise way or exercise your pedantic authority and reassure them that you plan to take up each of these questions in due time during future sessions but you feel that perhaps enough information has been covered for the first meeting without becoming confusing. If you are asked a technical question you do not know the answer to, the advocacy role of the IPT-ci therapist would have you suggest to the asker to let you research the question and then to get back to him or her with the answer.

Reading material can be recommended before departing, again with some judgment regarding whether it is directed at the patient or just the caregivers (if you conclude that the patient is too overwhelmed at this point to be expected to digest written materials). Classic texts such as *The 36-Hour Day* by Rabins and Mace or a variety of other available books and printouts are available. (See the list of references.)

When to See Family Members Separately

In the majority of instances, having the patient and family in the same room during the initial evaluation is a good idea for all parties. It gives the IPT-ci therapist the opportunity to engage the patient directly. It provides an opportunity to model the way you respectfully and patiently allow the patient to speak his or her mind (you are demonstrating respect for his or her autonomy) while simultaneously allowing you to gauge the reaction and stress level of the caregivers. In some circumstances, however, it is better to interview them separately. If the identified patient is paranoid of the motives of his or her loved ones or if he or she is too angry or hostile, then meeting both parties separately will allow you to assess the problems from both perspectives and then devise a plan for attacking the worst problems first and hopefully find a way for improving communication and begin building bridges to better mutual understanding. If the patient is disruptively paranoid, for example, the best strategy may be to try to administer appropriately dosed antipsychotic medication for a time until the paranoia becomes better controlled so that negotiations can then proceed on other fronts.

Giving the Patient the "Sick Role"

In IPT, when depressed patients cannot carry out their usual activities or duties due to low energy or low motivation, one strategy early in the engagement process of IPT is to give them the "sick role" to help alleviate any guilt they may be feeling for "not pulling their weight" or the self-perception of being "lazy." Comparing the illness of depression to a medical illness such as pneumonia may drive the point home. You can say something like "If you were in bed from pneumonia, you would not expect yourself to be raking leaves, right? Well, depression can be every bit as debilitating as pneumonia." With the added component of cognitive impairment, there may be real deficits that are impairing the patient's ability to function above and beyond the effects of depression alone. These may not be obvious at first, and oftentimes these deficits are not noticed by family members spontaneously. In these instances, complying with the IPT technique of assigning the sick role makes even more sense. It allows patients to be relieved of some duties they might find onerous and if family members are able to cover their duties, it sends a message of expected support. The relative contribution of depression versus cognitive impairment to the patient's waning function needs not be teased apart at the initial session; it will become clearer over time. As the depression is treated successfully, remaining decrements in function can be reasonably attributed to the cognitive impairment. It is advisable to triage the presenting problems into the most doable segment first and resist the temptation to present the "biggest picture" until you can be assured that all parties are ready to digest it. This will usually require more than one session to allow you adequate time to grasp all of the personalities involved and any potentially disparate agendas they may have.

Forming the IPT Contract

In the traditional IPT contract, the patient agrees to meet with the therapist on a weekly basis for 50 min, in face-to-face, individual sessions for 12 to 16 weeks of acute treatment. During those visits, the goals are to encourage agreement to work collaboratively, adopt the "sick role," complete the interpersonal inventory, establish a focus for the treatment, facilitate the expression of affect, test perceptions and new approaches between sessions, point out linkages between changes made in social roles and any reductions in depressive symptoms, discuss strategies for dealing with role disputes including role playing, communication analysis, and decision analysis, and then agree to terminate the therapy by the end of the agreed-upon time period. As patients with recurrent depression have been treated with IPT, a maintenance form has also been tested successfully, consisting of monthly refresher sessions to maintain gains achieved in the acute phase of the IPT and thus to hopefully prevent a recurrence of depression (most often in conjunction with a prescription for long-term antidepressant treatment). In the MTLD-1 study (Reynolds et al., 1999), those (cognitively intact) patients whose initial IPT focus was role dispute seemed to require the monthly maintenance IPT sessions (without an antidepressant) to avoid reverting to their old maladaptive ways and becoming demoralized and depressed again. Those whose original IPT focus was unresolved grief or role transition did not seem to need maintenance IPT to remain well (without antidepressant medication), probably because they adequately resolved their grief or role transition in the acute phase. Coadministration of antidepressant medication is used as indicated for recurrent depression in the real world and combination treatment is recognized to be the most efficacious approach in most cases of recurrent geriatric depression, particularly in those over age 70 (Reynolds et al., 1999).

In IPT-ci, the same format can be put forth; however, geriatric aged patients may have more logistical problems in keeping weekly appointments due to transportation problems, difficulty ambulating, or dependency on others who may not always be available. Flexibility in delivering IPT-ci is required. Of course, having gaps between sessions that are too long makes it difficult to continue the momentum of therapeutic progress that hopefully is taking place. The use of the telephone may be an adequate substitute to maintain continuity for those patients who can hear well enough and have adequate privacy to talk about sensitive topics when they cannot come in on a weekly basis. Home visits are another treatment option, if feasible.

As the engagement of family members in the therapeutic process is a key feature of IPT-ci, it is therefore imperative to stay in contact with them despite their other duties and responsibilities (with the expressed permission of the identified patient, of course). The IPT-ci modifications encourage the use of individual or joint sessions with either the patient or pertinent family members or caregivers. These meetings need not take place on the same day. What is important is that the IPT-ci therapist stay connected to the caregivers, particularly if there are worrisome behaviors that indicate executive dysfunction or if role disputes are the focus of the therapy that is being worked upon. In the real world of impediments to regular attendance, the goal for delivering IPT-ci should be placed on the accumulated number of sessions completed rather than a fixed number within a discrete period of elapsed time. Flexibility is key and the role of the IPT-ci therapist is to remind all parties of the "connecting threads" that link prior sessions regarding the problem area(s) identified as crucial to making progress, particularly for patients who have memory impairment or for those caregivers who seem to need a lot of basic education to understand what they are observing. In this electronic era, contact with caregivers might include the use of the cell phone, email, text messages, or faxed in reports to stay connected as needed. The flexible role of the IPT-ci therapist is to be the integrator of information from all available sources (including medical reports, imaging reports, neurospychological testing, legal proceedings, etc.) in order to create the most accurate composite picture of the current functioning of patients, to identify their immediate needs, to anticipate their future needs, and to advocate for their overall welfare. In addition, the IPT-ci therapist brings to bear the appropriate psychotherapeutic intervention whether on an individual or joint basis with either the patient or various involved caregivers. The use of printed goals for the problem areas being worked upon can help the identified patient with cognitive impairment to recall important points between sessions and it also may help galvanize caregivers to remember the role they agreed to play in assisting the identified patient like a coach.

Although the time frame of intervention for traditional IPT for depression is 12–16 weeks, adaptations of IPT for monthly maintenance sessions for preventing recurrences of depression (IPT-M) have been successfully implemented as mentioned earlier for 3 years in one controlled trial (Reynolds et al., 1999). In a similar vein, IPT-ci can be used flexibly as an acute short-term treatment to attempt to stabilize an identified patient with depression and/or cognitive impairment (with or without caregiver involvement as indicated) as well as being a format to follow the identified patient long term using a flexible schedule of follow-up sessions to best meet the needs of the patient and the caregivers. IPT-ci can have a flexible range of visit frequency, from weekly visits with a cognitively intact depressed elder who is essentially receiving traditional IPT to less frequent meetings for patients who must be driven by caregivers or who cannot make weekly appointments. Patients with predominant cognitive impairment who are undergoing a dementia work-up that might take some weeks to complete might best have a follow-up visit after the tests results are available, at which time the focus becomes the interpretation of the results and devising strategies for coping with their ramifications. The frequency of visits necessary to adequately educate the identified patient and caregivers may also vary according to their receptivity and the complexity of the clinical situation (whether there are concurrent medical problems being treated or whether cessation in driving must be negotiated immediately, for example). In the case of identified patients who have cognitive impairment, regular follow-up allows the IPT-ci therapist to monitor the status of the coping ability of both the patient and the caregiver(s) both of whom can experience changes in their lives that impact their ability to maintain the status quo. Most obviously, if the cognitive impairment is due to a progressive dementia, further decline is expected over time albeit with an uncertain trajectory. Caregivers can also experience their own changes in health or well-being, job stability, childcare concerns, or, in the case of a caregiving spouse, their own cognitive decline that can impact the "coping system" and require a reevaluation of available options or strategies to meet the new challenges optimally. One of the adaptations in IPT-ci is a shift from viewing the therapy as a short-term approach as in traditional IPT to viewing it as a long-term comprehensive management strategy for clinicians to use in managing older adults with depression, cognitive impairment, or their combination acutely. The goal of IPT-ci is for the patient to achieve a steady state during the initial treatment stage and then regular follow-up meetings are used for reevaluation and potential renegotiation of appropriate new strategies for the remaining lifetime of the identified patient.

8

Incorporating Family/Caregivers Into the Treatment Process From the First Meeting

We will define a caregiver as a person who takes an interest in the welfare of the identified patient when cognitive impairment and/or depression raises his or her concern. Typically, caregivers are adult children or spouses. Each of these groups has their own particular needs to take note of. Less commonly, the caregiver may be another relative, such as a sibling, grandchild, niece, or nephew, or trusted friend, neighbor, or church member. We will concentrate our discussion on the first two groups.

Why is considering the needs of the caregiver important? Obviously, if the patient is showing signs of growing dependency on the caregiver and the caregiver becomes unable to carry on with that role, then the patient may be at genuine immediate risk for the consequences of that loss. For example, during a follow-up visit with a depressed patient who also showed significant cognitive impairment, his accompanying wife experienced a heart attack in our waiting room and was rushed to an intensive care unit. By default, we were entrusted with the immediate responsibility to care for her husband who could not negotiate his own way home or care for himself. We called his son who lived out of state, and he made expedited travel arrangements and helped us secure temporary housing for his father with a trusted neighbor.

The identified patient's level of anxiety and depression can also worsen with perceived losses of support experienced by the caregiver. Beyond these concerns, the quality of the relationship between the patient and the caregiver may be improved upon with better understanding of the real cognitive deficits of the identified patient and any achievable improvement in depressive symptom resolution. If cognitive deficits are embraced as the "new reality" to be contended with by both the patient and the caregiver, then negotiation can begin to find more adaptive ways of coping with this new reality. Caregivers can experience relief, empathy, altruism, and rekindled kindness toward the patient when they understand what the patient can or cannot do and may subsequently redouble their efforts to help the patient to compensate as best he or she can. On the other hand, if caregivers do not understand the true deficits of the patient, they may proceed with misinterpretation, resentment, and even punitive reactions.

The Adult Child Caregiver

Let's consider adult child caregivers first. In most cases, they will be female, either daughters, daughter-in-laws, or granddaughters and less commonly sons or grandsons, nieces, or nephews. From a demographic point of view, adult caregivers usually have the dual stresses of caring for their own children's needs as well as those of their parents or grandparents. In today's society where both marriage partners usually work, juggling their nuclear family needs, their parent's needs, and their work responsibilities frequently requires them to multitask all day long, which can be exhausting and hard to sustain for the long term. On the strengths side of the equation, many adult child caregivers are often highly motivated, easily educated, fully capable of assisting their parents, and willing to incorporate suggestions from the IPT-ci therapist to implement plans for improving a stressful situation as the following case vignette will illustrate.

CASE VIGNETTE 15: DAUGHTER IS KEY TO UNLOCKING MOTHER'S STALLED RECOVERY

Doris, a 74-year-old married woman of three, was admitted to an inpatient geriatric psychiatry unit for severe depression and weight loss. She made a very slow recovery in the hospital and remained convinced that she could no longer do any cooking or baking upon her return home despite moderate improvement with antidepressant medication and ward milieu. Doris lived in a rural setting and prided herself on her culinary skills such that her children traditionally came every Sunday to enjoy the fruits of her labors. She stated that she no longer recalled any of her recipes and did not even want to venture into the kitchen as it demoralized her. No amount of counter arguments were effective. Doris was referred for outpatient follow-up and agreed to a course of IPT-ci. Her husband, a retired industrial technician, was attempting to do the cooking as Doris refused and was instead spending long periods of time in bed stating that it was the only place she felt comfortable. Her husband tried to cajole her to at least try to cook but she refused and as he grew more frustrated, he was noted to raise his voice toward her on more than one occasion. Furthermore, her husband had been accustomed to coming home to a ready dinner during his working years while his wife raised the children and managed the household, thus he was feeling stretched and out of his element to fully take over the cooking in the home.

After several meetings, it became clear that Doris suffered from executive dysfunction, a form of cognitive impairment which was made worse by her superimposed depression to the point where she could not see a pathway to improve her own function in a rehabilitative sense. Her husband was highly frustrated and losing patience. Her mood had improved on the medications compared to her status when she entered the hospital but she was clearly "stuck" in a marginally functioning state that did not produce any of the traditional gratification that she received from preparing food for the people she loved. She said he hated to get out of bed as she had no purpose in life.

As joint and individual meetings with Doris and her husband seemed to produce little progress to restore a higher level of function, we decided to invite her daughter Paula to come for a joint session. Paula lived nearby with her own small children and readily agreed to come to try to help in any way she could. After acknowledging her own frustration that her mother seemed like she was not progressing beyond the modicum of improvement noted in the hospital, she agreed to try a new approach. Since the task of food preparation seem too daunting for Doris as she complained she could no longer recall any recipes, her IPT-ci therapist suggested that Paula come over on an afternoon to make cookies and ask Doris to "help." She was instructed to place no demands beyond what Doris could clearly handle and if that meant she just mixed the batter, so be it. Paula's instructions afterward were to make sure that all cookie consumers were told to thank Doris for helping to make such delicious cookies. The idea worked, and Doris seemed a tiny bit more comfortable in her old kitchen although she remained pessimistic and reclusive. Paula was also instructed to be sure that her father noticed how she talked to and showed patience toward her struggling mother. This paid off later when Doris's husband took the literal hint and asked if it would be OK to try to convince Doris to go "with him" to their church cookie baking event. With some trepidation, Doris agreed and her husband took the cues learned from his daughter to pay attention to signs his wife was getting frustrated or tired and to jump in to help her or insist that they take frequent breaks. The camaraderie of baking alongside her fellow parishioners and the successful results further helped Doris to regain some lost confidence and she began to try to cook simple dishes again in her own kitchen.

As the Christmas holidays approached, Doris began to fret again that she could not conceive of doing all of her usual holiday baking that consisted of a dozen different varieties of Christmas cookies. After another joint meeting with her husband, Doris seemed to be able to adapt to her IPT-ci therapist's suggestion of limiting her contribution to the four most favored varieties, which she eventually agreed was a better fit with her current age, stamina level, and cognitive status. Slow progress continued over many months for Doris where she regained a significant portion of her prior repertoire of dishes and her mood improved to her old baseline even though everyone realized that her new cognitive capacity dictated that she needed to be reminded at times not to overdo her efforts as she still became easily exhausted and overwhelmed. After 2 years of follow-up, she remained on her antidepressant medication and a memory-enhancing agent. She and her husband attended follow-up sessions every 3 months, which seemed about right for them.

CASE VIGNETTE 15 (Continued)

Doris did not decompensate when her husband was diagnosed with a rare form of cancer and required a lengthy treatment protocol to fight it. She had by then taken over almost all of the kitchen duties again and said she tried to prepare foods that her husband liked to eat to help maintain his morale.

Case Vignette 15: Discussion

This vignette illustrates the additive demoralization that cognitive impairment and depression can bring when they coexist. The core of life's activity for Doris was her role in the kitchen and the gratification she derived from doing it well. Her confidence was shattered by her depression, and her cognitive impairment made it extremely difficult to chart a course, like a ship at sea, to work toward greater levels of function. This inability made her feel even more hopeless and useless. A new tact needed to be employed to try to convince Doris that she could work toward improved functioning (if not back to 100% of her peak). With some coaching, her daughter Paula successfully modeled an empathetic and patient, step-wise action plan to convince Doris that she still had considerable culinary skills inside of her and to set an example for Doris' husband to emulate. If you imagine Doris entering her deep depression by slipping down a spiral of increasing demoralization and perceived loss of functionality and purpose to a point of feeling paralyzed and useless, these interventions countered that state of mind by creating an upward spiral of a series of small successes that brought more gratification, improved self-esteem, more hopefulness, and restored function (with identified limitations).

Adult Caregiver Issues

Adult children can also have their own problems and sometimes they are part of the problem that precipitated the depression in their parent in the first place. Personality clashes, financial dependence, housing arrangements, old resentments, blended families with multiple allegiances, uncompensated mental illness, or drug and alcohol abuse can all complicate the parent–adult child relationship and contribute to the onset or maintenance of the patient's depression.

We have already established that although adult child caregivers are not the patient, their needs should also be assessed, at least informally, by the therapist merely being observant while evaluating the patient. Obvious clues to high levels of distress are behaviors such as breaking into tears of frustration, angrily correcting their parent, or dismissing the parent by speaking for him or her, or ignoring his or her input. The appearance of caregivers is also important: do they look tired, hassled, stressed, unkempt, and so on. Inquiring about how difficult it was to come with their parent for the visit can be an invitation to hear about their stress level. A casual inquiry about whether they needed to take off time from work and what kind of work they do can also be informative. Accompanying small children and cell phone use during the meeting can also be clues to their other responsibilities.

The quality of the communication between the adult child and their parent can give clues to their level of understanding of their parent's cognitive status. Do they take time to explain ambiguities? Are they patient? Usually by the time an appointment is made and the adult child accompanies their parent, the caregivers have already formed some opinions about whatever problems prompted the appointment, but not always. Also, the accompanying adult child may not be the main caregiver, he or she may be peripherally knowledgeable and just happened to be available to drive the patient to the appointment that day. The extent to which the main caregiver grasps the true extent of the dual problems of depression and cognitive impairment will be a barometer of the work ahead in IPT-ci and will dictate the extent of education required. Beyond that, the caregiver can help formulate an action plan that will help to alleviate depressive symptoms and clarify remaining cognitive abilities and alternative coping strategies in order to maximize quality of life.

If it becomes apparent that the accompanying adult child caregivers are overwhelmed or in need of specific services for themselves, it is of course warranted to suggest a referral for a specific kind of help be it mental health help, drug and alcohol treatment, legal counsel, or referral to a social service agency to try to obtain more hands-on helpful services. The IPT-ci therapist should bear in mind, however, that the identified patient is his or her primary responsibility and the caregiver is, as the detectives say, a person of interest but not a patient *per se*. If role disputes between the identified patient and the adult child caregiver become the focus of IPT-ci, then the advocacy must lean more heavily toward the patient. Hopefully, there are grounds for negotiation that can be facilitated by the IPT-ci therapist through either joint or single meetings with either party.

It is important to recognize that adult child caregivers may be going through their own role transition from whatever longstanding relationship was present premorbidly to one where they must rethink the needs of their parent in absolute terms. In this new relationship, they will need to provide more supervision and assistance and sometimes to confront their parent who is unwilling to comply with tasks or changes that appear to be very reasonable and in the parent's best interest for safety or well-being.

Keep in mind that adult child caregivers may be overwhelmed with other strained relationships that demand their attention such as acting out teenagers or strained marital relationships, the crux of which might be related to the time and energy they devote to the care of their parent as the following case vignette will illustrate.

CASE VIGNETTE 16: ADULT CHILD CAREGIVER VOWS TO "BE THERE" AT A HIGH COST

An 82-year-old man was brought for an evaluation by his daughter and son-in-law. The identified patient showed little desire to elaborate and seemed unsure of why he was there during the interview. The daughter did most the speaking for all of them. She told a story of bringing her father to live with her and her husband about 18 months earlier when her mother died. She stated several times that she wants to do what is best for him and revealed that she feels guilty that she could not "be there" for her mother before she died but is determined not make the same mistake again. She is an only child. She was tearful during the interview and appeared to be possibly suffering from depression herself. During the 20 minutes of telling her story, her husband sat silently and when asked if he had anything he wanted to add, his first utterance was that they had not had a vacation or a day off from caregiving duties in a year. Further inquiry along these lines indicated an intense undercurrent of resentment and tension in their marriage related to the stresses and strains of providing care for the identified patient in their home. After assessing the older gentleman's needs directly, a separate appointment was made to learn more about ways in which the couple was struggling to cope and to begin the exploration of options that might work better, such as possibly considering some respite care options. After two such meetings, it became apparent that their marriage was highly strained and had been so for a long time, pre-dating the arrival of the wife's father. When the couple acknowledged this fact, a discussion ensued about possible marital counseling, which was refused. The husband, however, did express an interest in pursuing individual counseling, which was facilitated with several referral choices. His wife declined to have an individual referral.

Case Vignette 16: Discussion

This brief vignette illustrates the fact that multiple agendas may be occurring simultaneously and the role of the IPT-ci therapist is to listen empathetically and be willing to make a judgment about the degree to which the caregiving role or roles are impacted by these multiple stressors. If these other stressors can be clarified, specific referrals for appropriate intervention can be made without becoming involved in providing the treatment directly. The progress of the outside intervention and its impact on the overall caregiving capacity of the unit (whether positive or negative) needs to be monitored along with direct assessment of the identified patient's status. The outcome in this vignette is uncertain as they chose to pursue alternative follow-up closer to their home.

The reasons why a given caregiver steps forward are multiple but commonly include the factors of being female, living nearby, or being the most organized of their siblings. Sometimes adult caregivers are motivated by guilt, family expectations, or just a sense of decency, as well as abiding love or altruism. Some may be looking for a way to exit another unpleasant life situation (like a job they want out of), or may hope to be disproportionately rewarded financially in the last will and testament of their parent. Mixed emotions can make caregiving more difficult, for example, in the case of past child abuse.

The IPT-ci therapist should remain open to direct and more subtle communications from caregivers that mixed emotions may be complicating their efforts. Validating caregivers' struggle can be extremely supportive for them and suggestions for ways in which they might seek out their own therapy as needed can be valuable advice for them. The IPT-ci therapist may take note of clues that point toward one of the above situations but avoid direct engagement around the issue, seeking instead to remain focused on the identified patient's particular needs unless the caregiver's own problems threaten the safety of the patient or if the issue is clearly key to the caregiver's ability to participate in the therapeutic process. If caregivers specifically divulge issues they struggle with that seem to warrant other help, the IPT-ci therapist should acknowledge these issues and begin to negotiate how they might seek their own help while simultaneously clarifying that IPT-ci therapists cannot take up the issues in depth themselves as their responsibility lies first and foremost with the patient's needs. A referral to a colleague within the same office might allow for simultaneous visits for both to take place. Collaboration between therapists can be even more helpful (with mutual permission of course).

Sharing Caregiving Duties

Some families rally to the cause of an elder in need of more help and call family meetings, forge resolutions for making changes, and willingly share in any care burden between eligible parties. A more common scenario, however, produces one designated person as the main caregiver and others, who may be just as capable, take a backseat. If you ask the attendant caregivers why this is so, they will tell you that their siblings have lots of excuses like they are too busy, they could not imagine doing a better job, or they never got along well with the afflicted parent anyway, and so on. More often, there is passivity on the part of the minor caregivers that requires them to be activated for a specific task by the dominant caregiver only to revert back to passivity when the task is completed, leaving the dominant caregivers feeling resentful or depleted and often concluding that it is just easier to do all the tasks themselves rather than to attempt to delegate further.

If sharing caregiving responsibilities is a divisive family issue, an offer to have a family meeting for all interested parties can be an opportunity for psychoeducation that can ultimately benefit the patient. In this meeting, the patient's true limitations should be methodically outlined and caregiving needs made explicit. The IPT-ci therapist may be looked upon for help in negotiating reasonable strategies for sharing the responsibility of caregiving. Sharing HIPAA compliant vignettes of success stories of other families you have worked with can generate a spirit of soliciting ideas, energy, and input from all who have a stake in the long-term welfare of the afflicted person.

Inadequate or Abusive Adult Child Caregivers

Adult child caregivers are sometimes willing or well intentioned but compromised themselves and ultimately unable to be of much help. Alcohol or drug abuse, severe personality pathology, low intelligence, or other mental or physical illness may impair their ability to be adequate caregivers. It is often the case, for example, that children who have some disability such that they do not emancipate from their parents and continue to live with them as they reach an older age become the de facto caregivers, however ill-equipped they may be. Oftentimes, the looming crisis of "what to do with the dependent adult child" is anticipated and other capable family members participate in planning alternate strategies for care. When adequate planning never takes place, if one parent dies and the remaining parent declines in cognitive ability with or without depression, the disabled adult child is often ill-prepared to cope. The afflicted parents may have been in collusion with the avoidance of alternative solutions for several possible reasons. They may have feared being alone or they value the help provided by the dependent adult child for housekeeping tasks as they grew older or sometimes, the bond of attachment is just very tight resulting from a lifetime of protectionism such that there is an over-enmeshment that they will not consider altering.

CASE VIGNETTE 17: WHEN THE CAREGIVER CANNOT PERFORM EFFECTIVELY DUE TO MENTAL ILLNESS

An adult child with schizophrenia lived his whole life with his parents and continued to cohabitate with his mother when his father died. His mother slowly deteriorated in cognitive function and apparently, over a period of some years, she ceased to remind and support her son to follow up for his own mental health visits such that he stopped taking his antipsychotic medication. Their story made the newspapers when the mother was discovered bound to the banister with newspapers under her. Her well-meaning but now psychotic adult son stated that she was incontinent and he could not convince her to use the toilet. He could think of no other solution than to confine her to one area of the house and put newspapers under her to keep her from soiling the rest of the house like you might do for an untrained pet animal. Fortunately, his mother made a good recovery with proper care on geriatric inpatient unit and her adult son was simultaneously admitted to a different psychiatric floor for treatment of his psychosis. Both were then eventually placed in separate supervised living settings.

Case Vignette 17: Discussion

This case is an extreme example of a near-catastrophic outcome for a dementing adult cared for by a mentally incompetent relative, which further illustrates the necessity to adequately evaluate the competency of the caregiver(s) by being alert for signs of danger and by systematically exploring how caregiving is being carried out and by whom. This case would certainly qualify as a case of elder abuse from the point of view of the patient not having her basic needs met and being punitively confined even though the perpetrator was not acting out of malice but rather from a confused, psychotic state.

This scenario is often observed where an adult child who has had problems becoming fully independent is currently living with a parent who develops cognitive impairment. This adult child, by virtue of the problems that have kept him or her at home, is not always the most resourceful and capable caregiver. Addressing the needs of the identified patient often requires as assessment of the needs of the adult child as well in these cases. The next vignette will illustrate willful abuse.

CASE VIGNETTE 18: ROLE TRANSITION AND ELDER ABUSE

A 77-year-old woman with cognitive decline and depression is brought by two daughters: the younger daughter (about 30 years of age) is currently living with the patient and the older one (about 40 years of age) lives in the vicinity. The daughter who lives with the identified patient had moved back home several months earlier after a break-up with her live-in boyfriend. It was curious to note that that most of the talking about care issues was dominated

CASE VIGNETTE 18 (*Continued*)

by the daughter who was not the one who lived with her mother. When the younger daughter was asked for her input directly, she seemed to agree in a noncommittal manner to whatever was being discussed. After carrying out the standard evaluation and needs assessment, a follow-up date was set for several weeks later due to transportation problems and impending holidays. Preceding the following visit, the older daughter called ahead to ask for a private meeting during which she revealed that she could no longer remain silent for her mother's sake. She went on to paint a picture of her younger sister as a severe alcoholic who was terrorizing her mother on a nightly basis with verbal tirades and who was manipulating her with threats and excuses for why she continued to need money from her (presumably for alcohol purchases). The older daughter, who had clearly mustered her own resolve to take action to protect her helpless mother, told how she had filed a protection-from-abuse petition with her local magistrate to have her sister removed from her mother's home.

Case Vignette 18: Discussion

Due to the patient's cognitive decline, she was unable to "grasp the big picture" in this instance. She knew that she and her younger daughter seemed to "argue a lot" but did not seem to link the arguments to her daughter's nightly inebriation. The identified patient seemed to focus more on the details of each argument, and she was unable to link them to a destructive pattern that she was victim to. She merely saw her daughter as having "fallen on hard times" and that "she had her own troubles" and saw herself as trying to be supportive to her daughter by helping her financially until she could "get back on her feet."

In this case vignette of putative verbal and financial abuse, definitive legal action was being taken by the healthier daughter who made the decision not to stand by as her mother was being abused. If the younger daughter were the only "caregiver" and suspicions of abuse eventually became clear to the IPT-ci therapist, alerting the appropriate adult protective service agency to investigate may have been required. It is noteworthy that, when the identified patient (the mother) was asked about her relationship with the alcoholic daughter, her impression was distorted by her cognitive state such that she misinterpreted the true picture and kept making excuses and accommodations instead of being able to see that she was being manipulated. Being at risk for the devious intentions of "designing individuals" is one of the legal tests that judges use to determine whether guardianship is indicated. It is not difficult to imagine how this manipulative and abusive situation could have gone on for a long time unnoticed and unchecked as it undoubtedly does in many other settings. The involved clinicians who work with cognitively impaired patients must remain vigilant for these scenarios when making assessments and keep their senses sharpened for picking up new details that might point to potential abuse.

Caregivers Who Are Spouses

For caregivers who are spouses, a different set of issues arises. Here are a list of questions that might be explored at an appropriate opportunity to help delineate past and present social roles and any changes that resulted from increasing cognitive impairment, depression, or both. Through direct questioning during the interpersonal inventory, it is helpful to know what was the nature of the premorbid marital relationship by asking one or more of the questions in Table 8.1.

Gauging the awareness or level of denial in the caregiving spouse regarding the patient's depression and/or cognitive impairment is essential to be able to grasp the big picture and to begin to formulate the most appropriate focus for the IPT-ci. The psychoeducation may need to begin with very basic concepts for spousal caregivers who show a poor understanding of what is obvious to other causal observers.

Table 8.1 Sample Questions for Probing and Understanding Spousal Relationships

Who was more dominant in decision making, and who was more acquiescent?
Did both spouses work? share child-rearing duties? household chores?
What was the quality of the marriage historically?
Were there periods of separation or threatened divorce?
Was there a history of substance abuse in either party historically? Currently?
What is the current relationship with children and grandchildren for each spouse?
Are there hobbies for each, and are they currently active in them?
Have they discussed or written down advanced directives?
Have they discussed or made a will?
Who currently does the driving, grocery shopping, cleaning, bill paying, and negotiating with repairmen?
Who keeps the social calendar?
Have any of the above roles been reversed or taken over by the other spouse out of necessity?

Obviously, soliciting the spouse's view of the depressive symptoms of the patient and his or her reaction to living with someone with those symptoms is very important. Living with a depressed or cognitively impaired partner can be very depressogenic for the spouse. If the depression is longstanding on the part of the identified patient, the spouse may be feeling exhausted, overwhelmed, or angry that he or she has had to devote so much of his or her energy to the needs of the other at the expense of his or her own. It is well documented that caregivers often do ignore their own health needs when they are caregivers. Concrete suggestions or encouragement to seek appropriate medical care is one duty of the IPT-ci therapist as there is no quicker way to precipitate a worsened crisis than to have the caregiver become incapacitated due to deteriorating health from neglect. One should also not assume that non-patient spouses are free of cognitive impairment themselves. You may be witnessing an aging couple who have compensated for each other's deficits reasonably well for a long time but are now crossing a line into decompensation as a unit. It is certainly possible that both marital partners can be in varying stages of cognitive decline with or without depression and neither may be in a good position to assist the IPT-ci therapist to develop and implement a workable treatment plan. In these situations, an appeal, with proper permissions, to adult children may be indicated for decision making, particularly if it appears that a move to an assisted living facility is necessary. Caregivers who clearly suffer from significant depression and/or cognitive impairment themselves deserve their own referral for thorough evaluation. The strategy for intervention with the original patient presenting for help must now be incorporated into an overarching plan to include both parties.

Encouraging social interaction and/or satisfying group participation for cognitively impaired individuals has traditionally included enrollment in senior centers, hobby clubs, or even hiring companions for several hours per day, which serves the dual purpose of providing appropriate cognitive stimulation for the patient as well as freeing up caregivers to take care of their own personal needs in the interim.

Soliciting outside help from any available or concerned family would be the obvious place to look for assistance and, if no one is available, social service agencies may need to become involved to plan for contingencies for the couples' best interest. Sometimes, adult children who do not live in the locale decide that it is their duty to travel to their parent's locale and assess what measures should be taken for the welfare of their parents. The IPT-ci therapist in these cases can play a pivotal role in such negotiations, some of which may need to take place by telephone for out-of-town parties of interest.

Asking spousal caregivers how they understand what is going on to explain the patient's presentation is an opportunity to quickly gauge whether their understanding is based upon myth, outdated concepts, or merely attributing the changes to aging. Allowing them to answer with their own ideas before launching into your own psychoeducational effort is a planned strategy as part of the assessment. For caregivers who are not grossly impaired but who are marginally capable, extra effort to assure adequate understanding and the consideration of some types of outside help (such as home-health aids for bathing or housecleaning) or increased supervision may be the most pragmatic solution for the short term.

Longstanding marital conflict may also become evident. New onset marital strain can result from the impatience, poor insight, self-centeredness, impulsivity, or disinhibition manifested by the partner with cognitive impairment, especially if he or she manifests executive dysfunction. In rare cases where the marital strain is so intense or the common ground for negotiation or compromise is so elusive, the only humanitarian thing to do (after consultation and consent from all concerned parties) is to separate the warring parties and have them live out their lives separately. The author has seen this become a necessity in a couple of cases over 20 years, particularly when disinhibition leads to physical violence.

In the more common case of a relatively intact spousal caregiver living with a cognitively impaired husband or wife, a similar approach is warranted as was described in the adult child caregiver section. Many such caregivers, by virtue of a lifetime of knowing their spouse well, are able to spontaneously adapt workable strategies for coping with the observable cognitive impairment in their spouse. They may still need to cajole their spouse to be compliant with healthcare visits, medication regimens or bathing, however. Unlike adult caregivers, spousal caregivers may have less reserve energy and may be confronting their own health problems or growing frailty simultaneously. Oppositional behavior, personality changes, repeating questions, loss of a "true companion" or loss of the "man (or woman) I married" due to the deteriorating effects of cognitive impairment, depression, or both are commonly heard complaints. Less commonly heard complaints include threatening behavior, nonreciprocal sexual demands, and embarrassing social faux paxs.

Adult Child/Intact Spouse Collaboration

Now that we have considered the individual situations of the adult caregiver and the spousal caregiver, let us briefly consider other possible scenarios. Very often, the caregiving challenge is a shared one with the most common scenario being a combined presentation of patient, spouse, and adult child where the spouse and patient live together and the adult child lives nearby but sometimes travels from a distant location to attend the meeting. The adult child is present because he or she recognizes that one or both parents are showing signs of declining function and wants to do what he or she can to understand "what is really going on" and, not having any prior experience, wants to know what his or her role should be to help. After the assessment and the short-term treatment plan is put in place (to perhaps check labs, gather records from other sources, or start antidepressant medication, if indicated), the day-to-day implementation of any treatment plan falls to the spouse to carry out with input or supervision from the adult child caregiver. This collaboration may work well and help to restore more of the identified patient's prior level of function. If the caregiving spouse is not able to grasp concepts as presented due to preoccupations with his or her own health or if he or she is also in an early state of cognitive decline, then the role of the adult caregiver looms larger as the person who must somehow convince the caregiving spouse to follow the treatment plan or who must take it upon himself or herself to implement the treatment plan directly. Frequently, the adult child caregiver has already begun various global caregiving roles such as helping with shopping, repairs, and accompaniment to doctor visits for both parents.

Nonfamily Caregivers

Nonfamily caregivers can be unmarried romantic partners or roommates, friends, clergy or their designates, neighbors, or hired help. Obviously the bonds of attachment may not be as strong as family bonds in these scenarios but may be all that is available. Alternatively, there may have been a breach of some kind between the patient and one or more family members and the nonfamily caregiver may have stepped in out of altruistic necessity. Nonfamily caregivers usually see their roles as more limited and with less latitude to operate within. For example, they may not want to get involved in financial matters such as paying bills but merely want to see that basic needs and dignified living conditions are maintained. If there are ongoing role disputes between the patient and other family members, this scenario often creates tension with the nonfamily caregiver as well. We have seen cases of incredible dedication from nonfamily caregivers performing all kinds of caregiving roles. The bottom line for the IPT-ci therapist is that an assessment must be made of all available potential caregivers who might share in the responsibility with some understanding of what motivates each person to be a caregiver in order to be able to make some judgment about issues of safety and reliability when crafting a treatment strategy for the identified patient. The systematic exploration of these themes falls under the IPT-ci task of completing the interpersonal inventory.

9

Specific IPT Foci in IPT-ci: Grief, Role Transition, Role Dispute, Interpersonal Deficits/Sensitivity

Typical Problem Areas for Older Patients

The four foci or problem areas of IPT are role transition, unresolved grieving, role dispute, and interpersonal deficit/sensitivity. Table 9.1 lists issues or statements that are typical ones heard from older patients with depression and/or cognitive impairment, which may be a starting point for implementing IPT-ci.

Table 9.1 Typical Problem Areas for Older Patients

Role transition:

- "I miss what I used to do"
- "I am not happy with what I am doing now"
- Limited ability to accept reality
- Successful aging versus feeling overwhelmed
- Mourning the loss of cognitive ability, physical abilities, overall health, youth, mobility, intimacy, sexual gratification or perceived financial security

Role Dispute:

- "I don't like being told I lose my temper too easily"
- "I can't remember things well so I guess I ask my spouse over and over"
- "I feel I am losing my sense of independence"
- "I have been a good driver my whole life and I don't see why I have to give it up."
- "I have enjoyed a cocktail before dinner since I got married and now they tell me I am not allowed to have one anymore."
- "I worked hard to earn my money, and I want to spend it the way I wish."

Table 9.1 (Continued)

Grief:

- Mourning the loss of someone who died can be complicated by executive dysfunction.
- Deficits that were compensated for by the late spouse will often become unmasked or fully visible for the first time.
- Secondary risks for deterioration such as declining nutrition, lack of maintenance or healthcare, and curbs on alcohol intake might exacerbate with the loss of the "organizing spouse."
- Depression often follows grief and the attempt to cope with the lost support provided by the late spouse can be overwhelming in the face of someone with cognitive impairment. Even suicide can seem like a logical choice for someone with cognitive impairment who has limited ability to generate multiple potential solutions to the new problems they are facing since losing their spouse. In our group's experience, there seems to be more of a problem for men to adjust to losing a spouse than for women in this current cohort of elders.
- Risks of poor judgment may lead to abrupt decisions with deleterious consequences such as selling a home too quickly or jumping into a new relationship out of fear of being alone, or sweeping financial decisions that are not thought through carefully without the previously supplied help from the late spouse.

The following case vignettes will illustrate one or more of the points listed in Table 9.1.

CASE VIGNETTE 19: ROLE DISPUTE OVER SON'S CHILD-REARING PRACTICES

Mr. Rosso, a 78-year-old retired government clerk who was known by his family to be a lover of good food and a generous giver of his time and energy, came for an evaluation for depression. He seemed to be happily retired for several years, and he spent most of his time in and around his home and neighborhood and in the company of his wife of 51 years and his grown children and grandchildren. His depression was helped a great deal by an antidepressant medication, and over a 6-month follow-up period, he seemed content and well adjusted again. His wife began to complain a year later, however, that he was repeating himself and that he would occasionally forget entire events that happened days earlier. He agreed to a neuropsychiatric evaluation, which showed a pattern of neuropsychological testing consistent with MCI-amnestic type and was placed on Aricept® by his primary care physician, which he said he thought helped his memory a little.

About 4 months later, he asked to make an appointment and his wife accompanied him. She reported that he had "embarrassed himself" in their son's home by berating his son about the way he raised his children. Mr. Rosso felt that his son's children would not listen to requests from their parents and had to be told over and over to complete necessary tasks like picking up their clothing. He felt that "the law should be laid down" and that it was wrong not to set up consequences for inaction after repeated requests. He blamed his son for "not putting his children before his job" and told him he was working too much to parent them adequately. This theme is one he said he had put forth many times diplomatically over the years but this time, according to his wife and the reaction he got from his son and daughter-in-law, he came across as dogmatic, red-faced, enraged, and "over-the top." In asking for his description of the event, Mr. Rosso was prepared to argue his point further as it seemed clear to him that something had to be done and that he felt an obligation to speak up and take his son to task. He reported that it had been bothering him for a while and was having trouble seeing why everyone was upset with him. He was, however, prepared to give the situation the benefit of the doubt that he was possibly overreacting since so many people were saying so.

This scenario is a typical example of executive dysfunction. As in most social interactions, the main issue is less about what is right or wrong in black-and-white terms but how to express oneself while also using judgment about the appropriateness of doing so. The right time and the right way of making a point involves timing, anticipating the impact that your statements might have, gauging "your place," or the justifiability of your position to speak out, and so on. In this case, Mr. Rosso clearly felt strongly about the themes behind his accusation, but he had previously used tact, humor, and flexibility to make his points, but he now came across like a bull in a china shop. He had lost his cool, and he was seeing the issues as black or white and not at all in the gray area. As noted in Chapter 6, executive dysfunction results from a decrease in the efficiency of inter-brain neural connections most commonly due to accumulating damage from a degenerative process like Alzheimer's disease or multiple strokes.

After hearing Mr. Rosso and his wife out in detail, the IPT-ci therapist tried to take the "explanatory" approach by educating both parties about executive dysfunction and the cognitive impairment that often accompanies memory loss. Cognitive changes were pointed out to sometimes affect judgment (knowing when and how to levy constructive criticism), abstract thinking (being able to imagine in real time how the unfolding interaction might look to a stranger walking into the room), empathy (being able to anticipate another's feelings and imagine how you would feel in a similar situation), and so on. Mr. Rosso and his wife could see the similarity to their

CASE VIGNETTE 19 (*Continued*)

own situation in the examples given. Since both parties agreed they wanted to avoid the pain of similar outbursts, alternate options for avoiding similar embarrassing moments were eagerly discussed. Since Mr. Rosso and his wife enjoyed a trusting and solid relationship, a strategy was suggested where he "check in" with her about those situations that were beginning to anger him before he "took matters into his own hands." They agreed that they could work together to "table" discussions of passionate nature regarding the behavior of others until they returned home where his wife agreed to "hear him out" and to gently remind him of her views, if different, and to explain why she thought the way she did. They both agreed that the IPT-ci session where the issues were "aired out" was very useful and helpful and they appreciated "working up a strategy" for addressing Mr. Rosso's concerns. Mr. Rosso was happy for the support and validation from his wife and expressed some relief that he could depend on her to "guide him" when he "started to see red." He also agreed to adjust his dose of his antidepressant medication upward to help him to feel in better self-control.

Case Vignette 19: Discussion

This case illustrates the altercations and strain on family interactions that can take place with waning executive dysfunction. Mr. Rosso was initially adamant about defending his position although not defiantly so. He was concerned about all the negative reaction to his words and was willing to defer to the "doctor's judgment." He and his wife both appreciated the frank assessment of the situation in a nonjudgmental way with a focus on problem solving to help avoid such unproductive altercations in the future. Placing the blame on Mr. Rosso's changed brain status, which was also responsible for his memory loss and not attributable to a flaw in his character, was also helpful.

An interesting point to note in this case is the onset of major depression that, in retrospect, seemed to be a harbinger of the recognizable cognitive dysfunction manifested in the following 6 months. This is a scenario the author has witnessed several times over 20 years in practice.

Mr. Rosso's wife was nonplussed at first with Mr. Rosso's dogmatic stance, which was totally out of character for him. She was very invested in learning all she could to "help him" and was fearful that restricted contacts with the grandchildren could result if she did not help him to avoid other embarrassing outbursts. Having the therapist model ways to use calm suggestions such as "How about we take a stroll together and talk these things over between us" taught her to act as a coach by helping her husband with his judgment lapses. Learning to see the scenario as a result of cognitive impairment (ED in this case) that was out of her husband's control allowed her to step in to help him, which, in turn, avoided more embarrassment for both of them.

To counter Mr. Rosso's demoralization reflected in his statement that he "really messed up," his IPT-ci therapist pointed out that as his wife agreed to help him (with her intact judgment), he might be called upon to reciprocate in other ways. For example, he might help her to negotiate getting to and from doctor visits with her ambulation problems due to painful arthritis. Note that sometimes using concrete metaphors for comparison such as these make understanding the big picture easier for someone who has difficulty with abstraction. This idea of a quid pro quo of helping the other partner with skills that were more intact in one than the other was well received by both of them and allowed Mr. Russo to "save face" that he was not seen as the only one with a problem.

After a series of eight regularly scheduled visits, the Rossos are now seen every 2 months for follow-up IPT-ci sessions and there have been no further reports to date of additional inappropriate outbursts.

Abnormal Grief in the Context of Cognitive Impairment

In IPT, grief refers to the emotional adjustment to the death of another person. Grieving the loss of pets, personal property, or health would be, by convention, considered to be a form of role transition. How does one differentiate normal and abnormal grieving?

The grieving process in a person with cognitive impairment can be exaggerated or distorted in a number of ways. In the case of a spouse who dies leaving a cognitively impaired partner, it is often the case that the true cognitive disabilities of the survivor quickly become more obvious to others as the extent to which the late spouse was compensating for his or her waning ability is no longer present. If practical issues such as bill payment, tax preparation, house repairs, yard work, and auto maintenance have been the purview of the late spouse, the prospect of needing to assume these new roles can be daunting even for a cognitively intact person, let alone one with cognitive impairment, who may experience panicky anxiety as a result. Conversely, those whose cognitive impairment is characterized by poor insight can seem oblivious to the tasks ahead that would induce some anxiety in any reasonable person. Usually, after a death has occurred, there is a reassessment of the current and future needs of the surviving spouse by other concerned family members and friends. The role of the IPT-ci therapist is to learn about all the options being considered such as possible moves closer to adult children, downsizing living arrangements, or formalizing support systems and then to help facilitate the discussion with the patient and other concerned family members to negotiate the most reasonable choices. When concerned family or friends are not available, exploration of available resources within the community may need to be pursued. A referral for social work services may be beneficial to this end.

It is usually preferable not to have the grieving patient be confronted with the strain of a move too soon after the death of a spouse has occurred as this adds the strain of the role transition of moving to the already existing strain of grieving.

Progress working through grief can also be distorted by cognitive impairment whose features might include decreased insight, less flexible thinking, and decreased problem-solving ability. This does not mean that cognitively impaired individuals will not work their way through the grieving process eventually but it may take longer and they may be more apt to get "stuck" in their grieving process. For example, cognitively normal individuals often get "stuck" working through mixed emotions involving the deceased when old hurts, resentments, infidelities, or unfulfilled dreams that engender feelings of anger, hostility, remorse, blame, or emptiness may need to be acknowledged and reworked in order to complete the grieving process. Some cognitively impaired individuals are at risk for chronic grieving or depression, by virtue of their declining cognitive ability, as they are less able to work through their mixed emotions, reassess their current life, and make reasonable plans for the future. This life review reprocessing involves making judgments about the past, present, and future that requires some ability for abstract thinking that may be impaired. The IPT-ci therapist may need to be more explicit, directive, and more willing to bring other concerned family into the treatment to shore up coping ability, objectify short- and long-term goals, and enlist support to take action when it seems to be difficult for the identified patient to do.

Traditional IPT asks the grieving patient to focus on the immediate time period right before and after the death as this time period is often highly charged with emotions, which are easily accessible for in-depth review during this period. In IPT-ci, this task may be more complicated as the events may be less well recollected and the final series of events may seem more blurred as the following case vignette will illustrate.

CASE VIGNETTE 20: GRIEVING DISTORTED BY COGNITIVE IMPAIRMENT

Barbara, age 74, was referred for psychiatric evaluation by her PCP who could find no cause for her multiple somatic complaints despite extensive investigation. She had been showing up in emergency rooms repeatedly with "severe pain" in multiple sites that did not seem to fit a known anatomical pattern or have an explanatory cause. Her primary care doctor thought she might be depressed. She also showed mild forgetfulness, poor insight, and illogical conclusions that were sometimes difficult to follow. Her premorbid functioning had been reasonably good as a homemaker and caretaker to her ailing husband for many years.

Further history of the onset of her multiple somatic complaints seemed to coincide with her husband being placed in a nursing home where he died 2 months later. Barbara described taking care of her husband for many years before reluctantly agreeing to his placement as suggested by his doctors when his multiple medical problems became too hard for her to manage. She described her husband as a domineering and selfish alcohol abuser who rarely considered her point of view in any decisions. She deferred to him on every matter that required a decision stating that she was afraid of him although he did not have a history of domestic violence toward her. She did say that he never showed any signs of affection for her either but quickly went on to say that he was a decent man to his children and someone who "never missed a day of work or a paycheck." When asked how she handled his transition to the nursing home, she replied that she was relieved that "he was getting along great with all those women" referring to the nursing home staff. She visited him regularly but did not feel that he was glad to see her or that it mattered if she visited or not. He was not asking to come home. When asked what she was doing differently with her own life since his death, she replied that she was now free to go to the mall or to shows with her daughters and that they were having a "great time."

The temporal relationship between the placement of her husband in the nursing home and his subsequent death and the onset of her multiple unexplainable pain complaints suggested that the two were possibly connected. Perhaps her symptoms were a form of grieving for her life of sacrifices that she was now painfully aware of? Perhaps she did not feel needed anymore? Perhaps she felt guilty on some level that she was relieved to be rid of her unloving husband?

Unfortunately, Barbara could not be made aware of any such connections despite exploratory efforts and repeatedly blamed her symptoms on getting the wrong drug from a previous doctor who treated her for stomach pain (an example of concrete versus abstract thinking). Her symptoms were quite severe and she continued to ask for pain relief and presented to emergency rooms at night when she could not stand the pain. MRI scans and other more esoteric diagnostic tests were invoked to rule out any potentially missed diagnoses but none were ever found. Her somatic preoccupation appeared to be of psychotic proportion with somatic delusions and several attempts were made to ameliorate her suffering with antipsychotic medication trials in addition to continued attempts to use IPT-ci. Her daughters were also growing more exasperated by their mother's clinging demeanor and loud complaints that no one believed her pain.

Case Vignette 20: Discussion

To succinctly sum up this complicated and difficult case that required many trials of medications and multiple attempts to engage the patient in treatment over 18 months, she finally made progress with an adequate dose of antipsychotic medication and was eventually able to make the statement: "I never realized how angry I was for all those years I waited on him." Her cognitive impairment made it difficult for her to use insight to put her resentments about her marriage into perspective without the efforts of her IPT-ci therapist to continually refocus her on the task of working through her resentments about her "wasted life."

As is frequently the case with grieving individuals, even those who are fully cognitively intact, working through their mixed or conflicted feelings about the deceased is often the major task that is inhibiting their progress. Helping Barbara to give voice to her anger and resentment about how her life was not as fulfilled as she would have wished while living with her husband, subsequently allowed the sadness that was also present but overshadowed by her anger to be acknowledged as well. Subsequent sessions focused on the "good aspects" of her husband, namely his consistency as a breadwinner, his unfailing devotion to his children, and for a life lived "according to principles," however rigidly he enforced them with Barbara. Helping her to work through her complicated grief brought about a reduction in somatic complaints and less feelings of guilt for now being able to go where she wanted without wondering, in the back of her mind, whether her late husband would approve.

Coping With Role Transitions When Cognitively Impaired

Retirement, moving from a primary residence, stopping driving, stopping sports participation, and quitting clubs, advisory boards, or other organized activities are also changes in social roles that would understandably be expected to require some adjustment, and perhaps some mourning of the loss of the old roles. Thus, the first task of traditional IPT is to help the patient mourn the loss of the old role. Perhaps there are new roles that can be assumed to help to replace the sense of loss of the old roles as the case vignette in a cognitively intact steelworker adjusting to retirement illustrated in case vignette 2 in Chapter 1.

Individuals who have cognitive impairment still need to grieve the loss of the old roles as in traditional IPT. Sometimes, however, one can see the opposite where individuals simply walk away from long-standing participation in an activity, which seems inexplicable to those around them. When they are asked why they abruptly stopped singing in the church choir or attending their regular card playing club, they may say that they merely lost interest or that it became too much trouble to attend. Upon closer inspection, however, one usually finds some aspect that became too challenging in light of their evolving cognitive impairment, like the need to learn new music for the choir or to remember which cards had already been played. The astute IPT-ci therapist will explore the reality enough to be convinced of its accuracy and then empathize knowingly that these old activities can sometimes seem too hard to continue without risking an embarrassing mistake. The abruptness of ceasing the activity also speaks to a lack of appreciation or insight into the effect their leaving might have on others which is another form of abstract thinking, that is, the ability to be able to imagine a scenario from someone else's point of view.

Executive dysfunction can have the effect of limiting one's ability to be able to generate multiple alternative solutions to problems and then "test" the relative merits of each one before concluding which is best to proceed with. Concrete suggestions of substitute activities that are similar but less complex and therefore more doable might be readily agreed to. These activities can serve to restore some sense of belonging as well as help to maintain what the famous psychologist Erik Erikson called "ego integrity vs. succumbing to overwhelming feelings of despair" (Erikson, 1950). (See Friedman, 1999; Hoare, 2002; Weiner, 1979; Welchman, 2000.)

Role Transition: The Need to Stop Driving

Some role transitions are thrust upon the patient such as a demand to stop driving when it becomes clear that the risk is too great to do so safely (Bouwman et al., 2007). The impact of this requirement is often a severe blow to self-esteem, particularly for men in the current cohort of elders who usually did the lion share of the driving in their household or relating to their employment.

When the diagnosis of frontotemporal dementia was elucidated in Oscar's case vignette in Chapter 6, it also became clear he needed to stop driving. He had already done so at night on his own. The following list of bullet points in Table 9.2 are ones that can be useful to try to finesse approaching the subject of driving while maximizing self-esteem and "letting them down as easy as possible."

Table 9.2 Strategies for Tactfully Arguing to Cease Driving

- Acknowledge that they have been driving for 40 years (or whichever) and were known as a safe driver.
- Point out their current age and the fact that many in their age group need to stop driving for a variety of safety reasons.
- Point out that many in their age group stop driving after having an accident but that it is far preferred to "retire their license" voluntarily before having that first accident. This would preserve their pristine

Table 9.2 (Continued)

- driving record and avoid having them being saddled with guilt if, God forbid, they caused an accident that injured someone (or themselves).
- As a detailed explanation of their true neuropsychological difficulties might be hard to grasp or prone to misinterpretation, try to keep it simple. State something like "It is common in your age group to begin having slowed reflexes and your mind's computer can slow down too [insert an example here of something they said such as 'you even admitted you get confused sometimes when there is a lot of traffic']. The problem I am worried about [you say] is not driving to the local convenience store for bread or to attend church services near your home, it is when you are confronted with a surprise situation where you have to act quickly and accurately, for example, when a child runs out into the street after a ball. People with your 'memory test pattern' can overreact impulsively and that is what can cause an accident. We don't want to wait for something like that to happen to you before you stop driving."
- Spend some time illuminating the alternatives for transportation such as buses, senior transport services, or car pooling with friends, and confirm that their spouse or adult children are willing to drive them places. Can they walk to any nearby desired destinations like a convenience stores or post office?
- If they seem reluctant, a driving retest can be broached to allow them to "prove the test givers wrong" even though sure failure is highly likely. This exercise sometimes shifts the blame for loss of driving privileges to the "authorities" and away from the IPT-ci therapist or family members. These tests are often offered as part of other rehabilitation services and there is usually a cost of a few hundred dollars required. If the driver flunks the testing, some agencies inform the state directly.
- As a last resort, all states have similar laws that require doctors to report individuals who have medical or psychiatric conditions that might interfere with their driving to protect the public safety. You can remind them that driving is not a right in any given state, it is a privilege that can be granted or denied as the state sees fit for the best interest of the public. One could also point out that you, the health care professional, could be liable. If you did not report them and an accident were to follow, you could be sued. Inform the patient that since their test results indicated a pattern of problems that are known to suggest "possibly impaired driving ability," you are required by law to report your findings to the state. You can try following this statement with an offer that you do not need to send in any such

Table 9.2 (Continued)

> reports if they voluntarily "retire" (avoid the term "surrender") their license. At the very least, try to get them to agree to suspend driving until the driving retest can be arranged if they insist on that path.
> - For those who feel deprived without a driver's license or argue that they need it for identification for air travel and so on, many states will issue an ID that looks similar to a license but does not come with driving privileges and is solely for identification purposes.
> - In some cases, the author has worked with family members who have taken it upon themselves to disable the family car by disconnecting the battery or similar action and then try to sell the ruse that the car is unusable and too expensive to repair or replace to bring about driving cessation.

CASE VIGNETTE 21: ROLE DISPUTE—PARENT–ADULT CHILD

A 75-year-old retired financially successful businessman we will call Raymond came accompanied by his son to "get advice" about his daughter who is estranged from the family. Raymond's daughter had documented psychiatric issues which Raymond desperately wanted to find ways to resolve before he died as he had played the role of family patriarch for his whole life and had provided funds to various family members throughout their lives to help them to be educated and to prosper. Unfortunately, the estranged daughter had mismanaged her allotted funds and Raymond was not about to help her financially any further without conditions attached which she found onerous and countered with accusations of unfairness and discrimination. Raymond wanted to go to his grave with the satisfaction that he had helped to launch each of his children and grandchildren with a tailored plan of financial assistance. The subtext, according to his son who accompanied him, was that Raymond was recently using questionable judgment by making too many costly remodeling plans, by driving too fast, and by being demanding, overly authoritative, and quick tempered. As these behaviors were rather drastic exaggerations of behaviors that had been glimpsed by family members in the past, it became clear that a change in his personality was taking place. Taken together with his forgetfulness, his rigidity in decision making, and his judgment called into question by his children, a pattern of executive dysfunction was apparent.

As these behaviors were exaggerations of his usual style, the question had to be raised as to whether his medical status (coronary artery disease)

CASE VIGNETTE 21 (*Continued*)

or any of his many prescribed medications might be playing a role in these new preoccupations. He had, in fact, been switched to a more stimulating antidepressant in the previous month. From an IPT-ci perspective, encouraging reasonable approaches to problem solving regarding the role dispute with his daughter was the identified patient's goal for treatment. A rough practical tally of available financial assets was a useful exercise for him (it might be considered too private a topic for others) to help him to see how he needed to prioritize the spending of his remaining assets. Reassurance that he earned the money and could allocate it as he liked was tempered with reminders of the realistic fact that he wanted to leave a legacy of sound financial footing for his family members (including his surviving spouse) and that it was possible to spend through his considerable assets if he didn't keep an eye on the big picture (here the IPT-ci therapist is "lending ego" to help him to stay in concert with his own stated goals). By virtue of the trusted relationship that had already been established, Raymond seemed more willing to listen to his IPT-ci therapist than to family members for whom he still felt "responsible for." Over several follow-up sessions, along with a reduction in his dose of stimulating antidepressant, Raymond became less demanding and less dogmatic and worked out a practical plan for spending his assets with more oversight from one of his children who was financially successful and who, it was pointed out, would likely be taking over these same decisions in the event that Raymond died from his multiple medical problems. Over the following year, there were two other "family crises" that Raymond sought out his IPT-ci therapist to be able to discuss and "check out his thinking process." Occasional joint meetings with concerned family or phone calls were valuable input for the IPT-ci therapist to retain a good grasp of the big picture and thus to be able to make reasonable suggestions regarding decisions that were pending at each point in time.

Case Vignette 21: Discussion

In traditional IPT with cognitively intact individuals, the goal for a role dispute focus is to help the patient weigh the pros and cons of the options before them that often involve mustering the courage to speak the truth, stand up for themselves, sometimes levy an ultimatum and lastly, to consider whether, for example, a marital relationship may be too damaged to continue and thus to consider whether dissolution is the best option. In IPT-ci with cognitively impaired persons, the options for change are usually more limited. Marital relationships, although far from perfect, may still be the best (or only) option for the identified patient who may have neither the practical wherewithal nor the finances to leave the relationship and live independently at this point in his or her life. In Raymond's case, his son and his IPT-ci therapist were both keenly aware that his self-esteem was wrapped up in his view of himself as the "patriarchal giver" to all of his progeny even though he was stymied about how to cope with his uncooperative daughter who had her own problems and a different agenda. As his daughter was unwilling to negotiate anything at this point, and there was ample evidence of a long-standing impasse in their "seeing eye to eye," the focus of IPT-ci shifted to trying to help Raymond to see that he had already been successful, to an overwhelming degree, with his overarching plan for his children and grandchildren and that he did his level best to do the same for his daughter as well. Helping Raymond to see his own effort as one to be proud of and to come to terms with the possibility that he may have no choice but to accept the bittersweet endpoint that there is no more he could do to win over his daughter to his way of thinking, seemed like the most humane course to pursue under the circumstances. Raymond's desperate attempts to be overly aggressive with his authoritarian stance were not working and, in fact, were backfiring against his own stated goals. Helping Raymond to see that his best option (and seemingly the only option) was to leave the next move to his daughter to make and hope that she "came around" before the final chapter of Raymond's life ended.

Interpersonal Deficit or Interpersonal Sensitivity

The strategy for this problem area of focus in traditional IPT attempts to encourage patients to expand their usually limited sphere of human interaction. The IPT therapist recognizes that these individuals have shown a lifelong pattern of difficulty establishing and maintaining mutually satisfying relationships. This category will typically include patients with significant personality disorder features or traits. The strategy for working with such patients in a short-term therapy is not to attempt to change their personality structure (this would be unrealistic) but to work with the therapeutic aspects of the IPT therapist–patient relationship as a proxy for helping the patient to rehearse ways in which to become more comfortable with greater social contacts and satisfying interpersonal exchanges with other people identified in their interpersonal inventory.

When considering the patient with a noticeable decline in cognitive capacity that was once intact, we would generally include interpersonal problems under the role transition problem focus when the net effect of their cognitive impairment (with or without coexisting depression) has caused changes in social roles such as needing to drop out of formerly pleasurable activities such as driving, playing cards when memory wanes, and so on. For patients who have shown a lifelong pattern of difficult or unsatisfying relationships that is further complicated by declining cognitive function or an exaggeration of prior maladaptive traits or behaviors, then interpersonal deficit may be the most appropriate focus for IPT-ci.

Patients with cognitive impairment with or without depression who fit best into this category often have a more limited interpersonal inventory to draw upon. Problems to anticipate may be that they are less willing to try new things like going to a senior center or to allow home health aids to offer assistance when needed. Sometimes, overzealous attempts to cajole patients into group social settings that they find anxiety provoking or are uncharacteristic outlets for them can backfire with the patient's resentment of such attempts being directed at the very person(s) trying to help them. In these cases, the IPT-ci therapist may need to broker a "pullback session" to try to help to renegotiate building back a more trusting relationship and then trying different tactics that begin with less drastic changes.

CASE VIGNETTE 22: INTERPERSONAL DEFICIT—NO INSIGHT INTO DECLINING FUNCTION FROM COGNITIVE IMPAIRMENT

Gloria, a young-looking 80-year-old woman, came with her daughter Jessica who had quit her job and moved back to Pittsburgh to attend to her mother's needs as she noted progressively worsening memory on her visits home and through her regular phone conversations. Jessica sacrificed a high-paying job to come home but stated simply that she was single and that her mother needed her. Jessica first insisted on living in an apartment nearby while she helped her mother to negotiate house repairs, taxes, doctor visits, and "catching up" on many other "neglected items." Gloria, a widow for 12 years, had been previously active and outgoing, and she enjoyed her daughter's company and said she could not thank her enough for all the help she gave her. As Gloria's memory and executive function declined even more with such events like inadvertently locking herself out of her home when she walked out to get the newspaper and calling her daughter in a state of hysterical panic, her daughter decided to move in to live with her.

Gloria was an art lover and painted regularly throughout her life as an outlet of creative expression. As her cognitive impairment worsened, however, despite trials of memory-enhancing medications, her interest in painting waned. Gloria also showed little if any insight into her cognitive deficits and stated that she exercised regularly and did not need help of any kind. Her daughter painted a different picture of her rarely venturing out, never exercising despite encouragement, and forgetting that she ate an hour earlier and wanting to eat again as examples that frustrated her.

After many months, Jessica realized that she needed more for herself than the role of full-time caregiver and landed her own part-time job, which eventually grew in duties and responsibilities. She hired a series of aides to be with her mother when she was working as she felt she could no longer trust her mother to be alone and not make a grave error in judgment. Gloria was reported to be furious every day that the hired help showed up at the house and she pleaded to be left alone. Jessica tried unsuccessfully to use logical arguments to convince her mother and was observed during office visits to show brief periods of demoralization and tearfulness at her failure to "get through" to her mother.

Case Vignette 22: Discussion

When denial or lack of insight are part of the presentation of individuals with cognitive impairment, it is reasonable to try to see if they can grasp the concepts being presented in the psychoeducation but, if they are unable to or refuse to look at them, a point of diminishing returns is reached. Gloria's daughter Jessica was making logical arguments that were grasped by everyone else but Gloria; however, Jessica could not yet see that they were futile to continue with. She was growing more frustrated as her work demands grew, and she knew she needed to convince her mother to have hired caregivers more frequently, not less so. Her mother remained very verbal, articulate, and opinionated with a keen eye for humor and all things artistic but she could not see that she had cognitive limitations that put her at risk for the potential negative consequences of errors in judgment (such as locking herself out of her own home while retrieving the newspaper and then panicking when she could not figure out how to solve the problem).

The first order of business in IPT-ci with Gloria and Jessica, after it became clear that there was no insight on Gloria's part to work with or build upon, was to arrange a meeting with her daughter alone to be able to openly discuss her concerns without her mother present to rigidly refute each point being made. This allowed Jessica the opportunity to vent her frustration, to be educated more accurately about the reality of her mother's deficits (memory impairment, very poor insight, and poor problem solving skills) and her remaining strengths (preserved artistic and linguistic ability), and to then work on strategies to maximize her strengths and her quality of life. It was not difficult for Jessica to accept the fact that her mother clearly had cognitive impairment which affected her decision making and functioning and that it was highly appropriate for her to be arranging paid caregivers and making decisions for her mother that were in her best interests. With the futility of her previous attempts to use logical arguments pointed out, a new approach was suggested for a trial run. Jessica agreed to try couching her requests in positive terms by avoiding confrontation regarding the paid caregivers by simply stating that they are coming for her (Jessica's) own peace of mind even though she is aware that her mother resents them. Accentuating the positive by encouraging her mother to try some painting, however rudimentary compared to her prior artworks, was also suggested to try to bolster her self-esteem and to capitalize on one of her strengths (her artistic talent).

In that Gloria has no insight into her deficits and therefore she had no complaints that she wanted to address, the bulk of the work in IPT-ci in this situation revolved around reframing the approach her daughter Jessica was taking in caring for her.

In a subsequent meeting, Jessica shared some of her own issues that revealed more about her own motivations. She said she had left home when she was of college age to escape what she called "a mother who wanted to control my life." She visited infrequently in the interim, married and divorced and when it became clear her mother was failing, she decided she was unhappy with her job anyway and would return home and care for her mother. Living together with her mother eventually did not work well due to her mother's excessive demands and complaints so she obtained her own apartment and a part-time job and hired caregivers to help her mother. She also described being acutely aware that time was running out to try to make up for the lost opportunity of not seeing her mother very much over the years since she had left home and she tearfully confided that she needed to feel that she was successfully bonding with her mother after years of distancing maneuvers. She tearfully described how thankful she was for the opportunity for them both to "know and forgive" each other. Jessica shared that she was seeing her own therapist to further work on her own issues.

By all reasonable accounts, Gloria appeared to be suffering from progressive cognitive impairment with very poor insight into that fact. She seemed to live her life in the moment without much reflection about where she had come from, why her life was the way it was, and what she wanted for her own future. Gloria's daughter Jessica was clearly going through her own multiple role transitions and was struggling with her wish to restore her mother to health and to try to fulfill her own desperate need to feel reunited after years of absence. Her mother was clearly well cared for, therefore the IPT-ci therapist's approach was to encourage Gloria and her daughter to have regular quality interactions but also to recommend that Jessica pursue her own career interests since "trying harder" to do more things for her mother than she was already doing was unlikely to bring her mother any more benefit. Jessica was encouraged to continue to work through her own issues in her own therapy as it seemed like the right thing to do for her, personally.

Interpersonal Deficit: Patients With Personality Disorders Who Develop Cognitive Impairment

Persons who have never coped very well are at risk for coping even worse when they develop cognitive impairment. Personality style, traits, or classic profile personality disorders may show exacerbation of difficult to handle behaviors such as acting out, and primitive defenses such as projection, clinging dependency, or hostility. The following discourse will consider some of the more common personality traits/disorders and how cognitive impairment can impact their presentation.

Dependent Personality Disorder

For elders who endure the loss of the object of their dependency (most commonly due to a death occurring), depression and a sense of hopelessness commonly follows. Searching for a replacement object often creates panic in the remaining family members who do not wish to fulfill that role. Alternatively, a painful learning curve of "firsts" will need to be confronted regarding those tasks that that were formerly done by the person upon whom they were depending and that they now must learn to do for themselves. This was illustrated by the case vignette of Millie in Chapter 1. Had Millie showed signs of cognitive impairment as well, her coping ability would no doubt been even more difficult.

Antisocial Personality Disorder

The effect of cognitive decline on individuals with antisocial personality can further weaken whatever inhibition there might have been that limited such activities as sexual indiscretion, stealing, hostility, or insensitive comments or acting out with physical violence. This personality disorder (PD) is, of course, overrepresented in the penal system. In general, the aging process does, however, seem to have a calming effect on the intensity of urges to act out in one way or another in cluster B personality pathology; however, caution is the better part of valor when there is a long history of antisocial acts in a person now showing signs of cognitive impairment. Caution should certainly be exercised with any person with cognitive impairment who is threatening physical violence, especially in the case of those who have a history of violent acts. Direct engagement with the identified patient might successfully center on seeking ways to achieve a give-and-take approach with others when he or she wants something rather than appealing to more internal principles of self-control that may never have been well developed. Sometimes, working with concerned family or caregivers about how they can best approach the problematic issues by limiting their risk for bad outcomes is the only option.

Narcissistic Personality Disorder

Aging itself, and cognitive decline in particular, can be perceived by an individual with narcissistic PD or traits as a fall from greatness, which can precipitate depression. The loss of specific abilities or the stature that accompanies them is seen by these individuals as a depressing loss. The tactic of the IPT-ci therapist with these patients is to look for ways in which to maintain or bolster self-esteem by focusing more so on their current capabilities rather their past ones. This will require more reiteration of current strengths or remaining abilities than with other patients. Reminiscences about past accomplishments or "glory days" can bring a temporary boost to flagging self-esteem as well as direct appeals to their narcissism (a language they can understand) by pointing out that their functioning, to you, appears to be in the "upper echelon" compared to others you have known of similar age who were similarly challenged at this stage of their life (with cognitive impairment). If they respond to this notion, you can then suggest working together to try to continue to remain among the most well adjusted for what they are facing and then segue to concrete actions they might consider for improving social functioning and pleasurable activities to combat the alternative of hopeless rumination and depressed mood.

Borderline Personality Disorder

Patients with this diagnosis usually mellow with age and show less intense affect and less acting out although there are exceptions. Borderline PD is a risk factor for an early death, on average, owing to the tendency of these individuals to take risks (substance abuse, driving fast, etc.) and the risk for completed suicide (although they show many more threats than actual attempts). Patients with borderline PD can be more demanding on family, medical, or nursing home staff as they show evidence of cognitive decline and have less compensatory options available to them. As with younger borderline patients, it is imperative to set limits, particularly upon what you are able and unable to do for them in your therapeutic efforts. As intense and shallow relationships with "all good or all bad" swings of allegiance toward significant others is a well-known pattern for these patients, the IPT-ci therapist must keep this in mind while trying to formulate a treatment strategy with the simple goal of maximizing social function, minimizing depressive symptoms, and promoting stability in interpersonal relationships, particularly those that are needed to help the patient function day to day. Although traditional IPT does not try to alter personality and avoids transference interpretations, with patients who have interpersonal deficit as their problem area to focus upon, the patient–therapist relationship is used to role play strategies for social interaction with the suggestion that they try similar strategies with those within their interpersonal inventory and then report back in the next session about how it went.

As tumultuous relationships are not uncommon in the lives of patients with borderline PD, when cognitive impairment is added to the picture, even less stability might be expected. This scenario might necessitate engagement of the significant others or caregivers to model for them ways in which to better communicate, set limits, and show steadfastness in the face of gyrations in "all good or all bad" attitudes. The IPT-ci therapist also helps caregivers to find strategies to avoid the tendency of these patients to "split" one staff member or caregiver against another, for example, by forging agreements ahead of time for how to handle certain situations among all caregivers or nursing home staff and then to present a "unified front" of consistency toward the identified patient.

Paranoid Personality Disorder

Historically, the term "paraphrenia" was used to designate paranoid ideation that occurred in the context of advancing age. As this term was rather nonspecific, it has been replaced in *DSM-IV* nomenclature with delusional disorder when paranoia reaches the severity level of a false belief or a delusion except in the context of a diagnosable dementia where the presence of delusions is merely a specifier for that diagnosis. Patients with paranoid PD have a lifelong pattern of suspiciousness and distrust of the motives of others that do not reach the delusional level. With advancing age, however, several factors can contribute to or exacerbate these tendencies. Vitamin B12 deficiency is a well-known cause of paranoid delusions and dementia from Alzheimer's disease can also be accompanied by the same.

Some patients, who commonly become paranoid of neighbors whom they feel are spying on them or plotting to hurt them or steal from them, will want to move to escape the danger. Sometimes this works. At others times, a new fear replaces the old. It should also be kept in mind that complaints of theft of belongings is most commonly due to memory loss in which they forgot where they moved a valued object or money and then complain that it was gone and look for a culprit to blame (usually a family member or neighbor). In a few cases during the author's career, however, such fears were well founded as it turned out that certain unscrupulous family members were, in fact, helping themselves to their dementing relative's belongings or cash, utilizing the perfect cover of the patient's poor memory or the ease with which the patient could be manipulated with a ruse. This, of course, is an example of elder abuse for which healthcare professionals in this line of work must remain vigilant for signs of.

Searching for and treating underlying causes of paranoia should be pursued by checking B12 blood levels and instituting replacement with vitamin B12 when low as well as considering a trial of low-dose antipsychotic medication, particularly when the patient is showing high distress levels or threatening drastic action against would-be perpetrators. Patients with paranoid tendencies to begin with are more apt to distort reality when they begin to show cognitive impairment as well. Working with families to understand what they are seeing and to respond supportively is also imperative. Sometimes family members ask if they should "play along" with the delusion and then take some action to remedy it such as arranging for a mock confrontation to "get to the bottom of the problem." The therapist should advise them that such ploys rarely work since paranoid delusions are usually not amenable to logic or counter arguments.

Schizoid and Schizotypal Personality Disorder

These patients have odd appearances, odd utterances, and odd ideas usually of a lifelong nature. Schizotypal PD patients appear to have more symptoms that approach a diagnostic picture of schizophrenia. As these patients age and need more assistance or become cognitively impaired, they may need to come into contact with others, such as caregiving staff, who are unfamiliar with the oddness of such patients, leading to further distrust and misunderstanding. These patients are often aloof, isolative, and preoccupied with their thoughts, studies, or tasks that they deem to be important to them. When these patients are forced out "into the open" by growing physical infirmity or cognitive impairment, their anxiety often rises and they will often be resistive to suggestions that require their participation. For example, if admitted to a personal care home or a nursing home, these patients would likely be very uncomfortable being required to attend group activities or to share a room with a roommate who wanted lots of personal engagement. Sometimes, these patients are perceived as oppositional as they often speak oddly or less fluidly than others, and impatient staff or caregivers can easily write them off as uncooperative. The task of the IPT-ci therapist is to try to engage such patients and find out what they are thinking about, however odd, and then try to form a liaison by showing some interest in the subject matter they revealed. Subsequently, looking for ways to address their concerns or interests by building bridges to other people using the knowledge gained from such interactions, the IPT-ci therapist can help to maximize their overall function and cooperativeness with daily routines or compliance with treatment regimens for physical or mental health. With worsening dementia, these patients are at high risk for psychosis, in the author's experience, and consideration for low-dose antipsychotic medication may be needed when this is recognized.

The next case vignette will illustrate a person with schizoid PD complicated by cognitive impairment.

CASE VIGNETTE 23: RECLUSIVE OLDER MAN SURVIVES SUICIDE ATTEMPT

A 70-year-old man we will call Harry was referred for outpatient follow-up after surviving a jump from a bridge in a suicide attempt. He was treated, stabilized, and released from a psychiatric unit and court-ordered to have outpatient follow-up. His stated reason for trying to kill himself was that he was told his lease for his apartment would not be renewed and he could think of no alternative plan of action. He was retired from his job as custodian for a church for which he mostly worked alone at night. He had no contacts with family for years and no friends. He spent his free time watching TV. He spoke with odd syntax and was overly matter of fact: for example, he said he came because he was told he had to, and he did not seem to ponder very deeply that he had survived a suicide attempt. He professed neither gratitude nor remorse for having been rescued.

What became clear is that his limited social sphere, which mainly consisted of saying hello to his apartment neighbors, appeared to him to be shattered when it was announced his apartment building was converting to condominiums and he had to leave.

The addition of cognitive impairment to his lifelong schizoid PD set the stage for him to conclude he had no other alternative then suicide.

As the social service team on the inpatient unit had secured him a new subsidized apartment, his equanimity was again restored and he resumed his former life that was comfortable to him with a minimum of human contact. He came regularly for follow-up visits and did not appear to be depressed or hopeless any longer. Curiously, he never seemed particularly depressed even when suicidal; his action appeared to be more like a rational decision to him.

Obsessive-Compulsive Personality Disorder

Patients with obsessive-compulsive personality disorder (OCPD) can be very difficult to manage even without cognitive impairment. They are to be distinguished from patients with obsessive-compulsive disorder (OCD) who must show either obsessions, compulsions, or both that interfere with their social or occupational functioning to meet diagnostic criteria. Individuals with OCPD tend to be rigid in their thinking and compulsive in their determination with little room for outside influence. In my experience, they tend to be "yes, but" patients who seem willing to consider what you suggest might be helpful to them but then they return to say that they either did not try the suggestion or it did not work. Such patients often want to return to previous failed treatments (such as repetitive trials of once-failed medications) as they second guess that perhaps not enough time had been given to the treatment originally. For patients whose OCPD traits may have been adaptive in their peak years by allowing them to perform certain jobs well that required obsessional attention to details, the prospect of facing declining health, infirmity, or cognitive impairment often precipitates profound depression. When faced with a problem or adversity in their younger years, these individuals coped by intensifying their effort to come up with a solution, often with intense rumination, until the problem was conquered. When faced with a problem such as cancer, limited ambulation, or the awareness of cognitive decline that they cannot fix despite whatever extra effort is expended, they become very depressed as their main coping mechanism has failed them. These patients can be quite desperate in these circumstances and be a high risk for suicide given their rigid thinking that does not readily allow for alternative ways of looking at their struggle.

The task for the IPT-ci in these cases is first to review their accomplishments and abilities that allowed them to achieve their past success and, just like as in the case of narcissistic patients, to ask how they can reasonably accept what cannot be changed and focus their energy on that which they can still work on. Sometimes, the introduction of a new hobby that they can immerse themselves in (but is not so technical that it is difficult for those with cognitive impairment to carry out) can give them a renewed and reframed focus upon which to concentrate their energy. For one isolated man with OCPD who lived alone and complained of loneliness and boredom after retiring from his job as an accountant due to his increasing memory problems, a suggestion was made that he join a service club such as the Lions or Rotary Club, which he did successfully. This new endeavor served the dual purpose of giving him a social outlet and a sense of purpose in the charitable component that is the mandate of this type of organization.

Addressing Alcohol, Other Illicit Drug Use, and OTC Meds

The cognitive decline due to the early stages of a dementing disease such as Alzheimer's disease or dementia due to multiple strokes can certainly be worsened by adding mind-altering drugs. Let's consider alcohol first.

Chronic alcoholism, in particular, is a more difficult problem. When individuals who have a long pattern of regular drinking begin to have cognitive problems, they usually show worse behavioral effects of heavy drinking, which can include an increased risk for delirium, falls, motor vehicle accidents, sexual indiscretion, and physical violence. Depression often accompanies alcohol abuse due to the direct effects on the brain in addition to the realization, in some, that drinking is causing deleterious consequences. Some cities have alcohol rehabilitation facilities especially tailored for older patients but most do not. Encouraging sobriety and considering a family intervention with the help of the IPT-ci therapist to get the addicted person into treatment may be indicated. One memorable patient with chronic alcoholism was brought for regular appointments by his high-functioning son who remained compassionate toward his addicted father even though his mother had divorced his father. The patient became alcohol-free for the first time in his adult life when he developed cerebellar degeneration from the degenerative effects of chronic alcoholism and could no longer walk such that he was placed in a nursing home. Without access to alcohol, his son observed with some amazement, how kind, affable, and engaging his father was with the nursing staff and other patients. He quickly become a favorite of the nursing staff who enlisted his help for minor clerical duties and the quality of the relationship with his son changed from one of compassionate supervision to a true give-and-take reciprocal one for the first time in his son's life.

Alcohol use is certainly grounded in modern society as a source of enjoyment, a social lubricant, an accompaniment to fine dining, and even as a heart-healthy ingestible in the case of red wine (in moderate doses). Alcohol is part of a great many social settings and it is a popular pastime particularly for the retirees who have more leisure time to enjoy it. On the other hand, addictions to alcohol and other drugs is a topic that merits a whole book itself and there is no age limitation to its potential devastating effects. For our purposes here, it is important to point out that the effects of alcohol are heightened in later life when the older persons' ability to metabolize alcohol slows. Even for social drinkers, when cognitive impairment begins to be evident as well, a common scenario is seen where the afflicted individual consumes alcohol in the same habitual pattern that they have been following for years (or decades) but its effects are now different, namely, more slurring of speech, more uncoordinated walking, and more disinhibited behavior such as socially inappropriate or embarrassing language, gestures, jokes or blunt, insensitive comments or over flirtatiousness. When concerned family members recognize such a pattern, it is often difficult for them to know how to alter it, particularly if the individual denies there is an issue or reacts angrily.

As mental health professionals, we have the unique opportunity to impact this situation. Sometimes, pointing out the science behind the brain changes from chronic alcohol can be impressive enough to raise concerns in some people. Sometimes an authoritarian stance is successful. Sometimes some finesse can bring about the desired result in the face of declining cognitive ability; for example, suggest to caregivers that nonalcoholic beer be purchased and offered for consumption or suggest that caregivers offer to mix the drinks and keep them lighter in alcohol and sometimes even omit the alcohol altogether. Presenting the patient with a mixed drink that has other strong flavors, presented is a social setting that is traditionally conducive to social intercourse wherein the alcohol content may not be missed, would in essence involve taking advantage of the lowered insight and awareness that often accompanies cognitive impairment. The ambiance of festivity at social gatherings is often infectious such that a person with cognitive impairment will be caught up in the moment as well and will not require the noticeable effect of the alcohol to feel relaxed.

Episodic heavy drinking can quickly derail the stability of an older person who is developing cognitive impairment as the following case vignette will illustrate.

CASE VIGNETTE 24: ROLE DISPUTE WITH ALCOHOL-INDUCED DELIRIUM—JOINT INTERVENTION

Geraldine is an 84-year-old widowed African American who lives alone in her own home. She has one daughter who lives in a distant city, a plane-ride away. Geraldine was seen for treatment of her recurrent depression and had responded well to medication in our geriatric clinic and had been stable for several months. A joint intervention became necessary after a series of events unfolded. Geraldine's daughter received a call from a neighbor of her mother who reported that her mother had not been seen with the usual frequency and when this neighbor tried to engage Geraldine, she was abrupt and appeared to be unkempt and even paranoid. Her daughter flew home and called us after finding her mother in a state of poor hygiene, dehydration, and paranoia to the point of being fearful that her neighbors were going to break in and rob her. It further became clear that Geraldine had stopped not only her depression medication but also her blood pressure and cardiac medications. The key to her drastic drop in self-care appeared to be related to binge drinking alcohol. Geraldine had been known to overindulge with alcohol at various points in her past but had not done so for some years. She freely admitted that the approaching holiday season had made her nostalgic for her late husband and that she had purchased a bottle of bourbon to try and seek some solace from the pain of missing him.

To provide for Geraldine's safety in the short term, her daughter coordinated doctor-supervised interventions to see that she was properly hydrated, had all of her medications restarted as well as to monitor her for any signs of alcohol withdrawal. Geraldine fortunately made a rapid recovery and ceased to show any paranoia after a few days of hydration, good nutrition, rest, and proper medication. Her paranoia clearly seemed to be related to her binge drinking, perhaps in combination with underlying cognitive impairment manifested by executive dysfunction.

A joint meeting with Geraldine and her daughter a week later explored the events retrospectively in some detail with the stated goal of figuring out how to prevent a similar occurrence again. Her daughter was understandably frustrated, even exasperated, that she had to learn from neighbors that her mother had deteriorated dangerously, and she vented some anger toward her mother that she had brought it on herself by drinking alcohol to excess and leaving her daughter no clue of her intentions. Her daughter wondered out loud whether her mother might need to move to assisted living to avoid another occurrence. Geraldine voiced her regret that she had "overdone it" but did not seem to appreciate the gravity of the situation. She resisted attempts to discuss removing all alcohol from the house, interrupting her daughter on several occasions to say that she insisted on having alcohol in her home to offer to visitors and that she did not intend to drink anymore herself.

In fact, alcohol had been present for years without a pattern of overuse, but the tone of Geraldine's statements focused on her view that "a little alcohol never hurt anyone" and that she loved her home and had no intention of leaving it. Clearly, Geraldine had a legacy of a strong will and a matriarchal family rank, and she intended to continue to exercise it and not be limited, dictated to, or chastised. Geraldine's daughter shot a glance at the IPT-ci therapist with a slight shrug as if to say "see what I am up against?"

Case Vignette 24: Discussion

From an IPT-ci perspective, despite her strong will and determination, Geraldine had clearly been at great risk from her poor judgment. As she did not seem to take the series of events seriously, her executive dysfunction and denial indicated some continued risk. Fortunately, there had not been a recent pattern of binge drinking in the prior several years, and the trigger seemed clearly to be the anniversary of her husband's death which had now passed. When not consuming alcohol, her day-to-day functioning was more than adequate to live independently. The question remained though, was her poor insight due to her executive dysfunction a continued risk for other risky behavior that could again threaten her health and potentially her life?

Several phone conversations followed with her daughter to reassure all parties that a renewed level of stability was in fact being maintained as well as to further discuss ways in which more checks and balances could be put into place to trigger an early warning system if similar problems arose again. Due to job constraints, relocation closer to her mother was not an option, and Geraldine had stated her preference to remain in her home and "be left alone." Geraldine's obstinacy and lack of insight seemed to limit the extent to which further negotiations would be of further benefit.

A separate meeting with Geraldine's daughter was held to discuss options. With her Mother's renewed stabilization, it was agreed that it did not seem necessary to invoke adult services to force Geraldine to move into a more supportive environment. This would be the last resort if efforts to guarantee her safety were refused with a continued threat of dangerous decompensation. She had refused any offer of drug and alcohol services. Geraldine's attempts to elicit more "checking in" by various concerned persons as a way to keep better tabs on her and any risk for another episode of binge drinking were discussed. In the end, it was decided that the best course of action was to allow her to continue to live independently but try to watch her more closely for the time being.

Geraldine has been and remains a determined and opinionated woman. These qualities may have served her well through various adverse events in her life. Her depression had responded to treatment, but her alcohol binge could have been fatal. Her minimization of the seriousness of the event and the toxic effects that alcohol could still have on her indicates a continued level of high risk due to her executive dysfunction. Without the ED, she may have had the episode of binge drinking but perhaps not to the degree that led to severe dysfunction in her health. It was the combination of the ED and the alcohol binging that was so deleterious. The risk of poor judgment remains and therefore, external efforts to encourage healthy behaviors, discourage unhealthy ones, and maintain surveillance for possible lapses is critical going forward. Forging a strategy that is workable but also respects the dignity and wishes of Geraldine is the challenge for the IPT-ci therapist to help facilitate. The IPT-ci strategy for Geraldine (and her daughter) is a dual one. Encouraging Geraldine herself to be compliant with all aspects of treatment, abstinent from alcohol, and open to the good intentions of her daughter is one task. The other is working with her daughter directly to help her to understand that her mother's refusal to fully cooperate is a combination of her strong will as well as the effects of the executive dysfunction that limits her mother's ability to see the true gravity of the situation. Helping to maintain surveillance with regular visits and liaisons with other healthcare providers and safety net persons such as concerned neighbors was also discussed as a prudent strategy to maximize Geraldine's long-term quality of life.

The management of comorbid psychiatric disorders in addition to depression such as generalized anxiety disorder, panic disorder, or obsessive-compulsive disorder can be expected to be further complicated with the onset of cognitive impairment, as the following case vignette will illustrate.

CASE VIGNETTE 25: ROLE TRANSITION—COPING WITH LIFELONG OBSESSIVE-COMPULSIVE DISORDER DETERIORATES WITH WORSENING COGNITIVE IMPAIRMENT

Mr. Phillip, age 72, was brought by his wife for treatment for depression and "nervousness." He was a retired clerk. His wife still worked, leaving him to fend for himself during the day. He is an avid fisherman in his spare time, and both he and his wife were old-movie buffs. Mr. Phillip complained bitterly of being unable to sleep at night, feeling keyed up and anxious, and also depressed that could not seem to "pull himself out of this slump." After a careful history, he described intrusive thoughts that he could not let go of and that kept him awake at night. For example, if he could not recall the name of a movie star he saw on television he could not rest or go to sleep until he researched it on the Internet. This happened repeatedly and with greater urgency over time. He would sometimes get out of bed during the night with an obsessional thought and go back to the computer even though he knew that would mean another ruined night of sleep. These intrusive thoughts and compulsive behaviors met criteria for obsessive-compulsive disorder, which had never been diagnosed earlier in his life (probably because it had not been so obvious until recently). Mr. Phillip's anxiety was very intense and his wife was very worried. We discussed inpatient psychiatric hospitalization as the prescribed benzodiazepines and high-dose SSRI medication did not seem to be providing relief fast enough, even though he was showing some signs of oversedation. The addition of an atypical neuroleptic medication at bedtime seemed to help to improve his sleep. Over a period of 3 months of regular follow-up, he slowly made progress (which is typical for the drug treatment of OCD with high dose SSRI medication). He was also referred to an OCD specialist for a course of cognitive-behavioral therapy (CBT) to help him to learn to catch himself and avoid going down a repeated path of compulsive research whenever he could not recall the name of a movie star or and old movie. He eventually was able to learn to resist such impulses and "ignore them."

Mr. Phillip had not been previously diagnosed with OCD although he had been known as fastidious at work and in his personal habits. After his OCD symptoms began to respond to intensive treatment, his accompanying wife began to complain about "other problems" namely, his anger outbursts when she would point out to him that he had forgotten recent events or things she had told him. Mr. Phillip would argue with his wife that he was correct although when his wife collected a series of verifiable incidents he had, in fact, forgotten, he began to admit that maybe she was right that his memory was failing somewhat. With the high-dose SSRI treatment, however, he seemed to lose the intensity of caring about the consequences of the memory loss as much. He was greatly relieved to be able to sleep every night. His wife was also relieved that he was not locked into obsession-driven compulsive research at all hours of the night and was glad he appeared to be more relaxed, but she voiced continued concern about his memory problems. Mr. Phillip agreed to have further evaluation of his memory problems.

The best way to understand this case, in the author's view, is to understand that Mr. Phillip possessed lifelong mild OCD that was not disruptive to his work or home life overall and therefore it never brought him to treatment. The onset of memory loss, however, triggered a cascade of obsessional rumination, anxiety, and compulsive research to try to compensate. Sleep deprivation, frustration, intense anxiety, and depression resulted as the vicious cycle that showed no sign of letting up. The targeted treatment (psychotropic medications and a course of cognitive-behavioral therapy) for the OCD brought slow but definite relief of symptoms and the SSRI effect seemed to make him care less intensely about his predicament. After 6 months, he reported that he was now going fishing again by himself and was able to resist the occasional obsessional thought that he should research an answer he could not readily recall.

Case Vignette 25: Discussion

The IPT-ci approach in this case was to seek optimal treatment for the patient's severe anxiety and depressive symptoms, understand that their origin was linked to his declining cognitive ability (which had exacerbated his mild lifelong OCD), and encourage coping strategies to limit his anxiety and depression. Eventually, Mr. Phillip agreed to a work-up of his memory complaints for a possible burgeoning dementing disorder (he was subsequently diagnosed with early Parkinson's disease, which can have a cognitive slowing component in addition to tremors and slowed movements). He was started on anti-parkinson medication carbidopa (Sinemet®) under the care of a neurologist.

Mr. Phillip's wife accompanied him to every visit and was instrumental in guaranteeing his compliance with medications and follow-up appointments. She too needed to be educated about the connection between the onset of his OCD symptoms and the evidence of new onset memory problems. Mr. Phillip's symptom severity seem to be greatly lessened on his current medication regimen although time will tell what further impact any additional decline in memory or executive functioning might have on his future coping ability. Long-term regular follow-up visits are planned to try to maximize stability, coping ability and to continue to educate his wife about the most effective ways she can understand what is happening and intervene appropriately.

Just as traditional IPT does not promise to be an effective psychotherapy for all situations, recommending a course of CBT for Mr. Phillip is entirely appropriate as it is an evidence-based treatment for OCD with proven efficacy for building skills to gain better control of intrusive thoughts and not be held hostage to obsessions and compulsions. Nesting the brief course of CBT within the long-term strategy of IPT-ci, in addition to the appropriate psychotropic medication, helped to restore Mr. Phillip to an approximation of his baseline level of functioning.

The next vignette will illustrate the deleterious effect of improperly used prescribed medication in the context of cognitive impairment.

CASE VIGNETTE 26: ROLE DISPUTE WITH HUSBAND COMPLICATED BY MISUSE OF PRESCRIBED MEDICATION

Mildred, age 70, was brought to the clinic by her husband, a retired schoolteacher, for treatment of depression. Mildred was a lifelong homemaker who, in recent years, had suffered from increasing hip pain and unsteady gait complicated by her obesity. Mildred also suffered from comorbid anxiety symptoms and had been treated with benzodiazepines over the years by her primary care physician. Her husband reported that his wife had been frequently angry and critical toward him in recent months such that they spent long periods of time in separate rooms watching different programs on separate televisions. When asked about her marriage in a one-to-one session with her IPT-ci therapist, Mildred replied that he "never says one word to me all day long, not even goodnight or good morning; he is the reason I'm depressed." During inquiries about her overall activities, a pattern of increasing withdrawal from friends and more complex activities such as stopping card playing with friends became evident. Mildred could only recall one of the three items on the MMSE. She complained of middling insomnia, but her husband reported that she often took naps for several hours during the day. Eventually, it became clear that Mildred was also self-medicating inappropriately with extra doses of benzodiazepines when "she felt stressed." She had difficulty recalling when or how much of these sedating medications had been consumed (physiologically, benzodiazepines can contribute to cognitive slowing, disinhibition, confusion, and depression as well as incoordination, and increased risk for falls).

In an individual meeting with Mildred's husband, his view of their marriage was one that had not changed appreciably over the years. He felt they had been communicating on a regular basis. He described his wife's reluctance to socialize or even go out of the house much as an evolving pattern over the past several years but also voiced that this had not bothered him that much as he himself was more of a "stay at home type." He said that they rarely argued but that his wife did seem to have a "short fuse" and would isolate in another room when she was angry with him. When it was pointed out that Mildred appeared to have some cognitive impairment and that, in our judgment, it was not safe for her to self-administer her own medications any longer, he seemed somewhat surprised as she had been a frequent "doctor goer" and "always seemed to be trying some medication or other." He readily agreed, however, to carry out medication management in her best interest and was eager to try and understand her impaired cognition. When the disinhibiting effects of self-medicating with benzodiazepines were described as having potential effects on disrupting her sleep pattern (excess daytime sedation and disrupted night-time sleep cycle) as well as potentially causing a tendency to overreact to emotional distress, Mildred's husband was eager to listen and better understand the changes he had been observing in Mildred.

A plan was implemented with Mildred's permission to have her husband manage her medications, and to slowly taper her benzodiazepine medications and replace it with an SSRI that has antianxiety as well as antidepressant properties but is less apt to cause further cognitive slowing. From an IPT-ci perspective, a focus on role transition to a less mobile and self-perceived physically impaired person was made with a secondary focus on her perceived role dispute with her husband.

Mildred showed steady progress, a brightened affect, and reported less depressive symptoms over the ensuing weeks. She did not object to her husband taking over her medication dispersal; on the contrary, she relished the extra attention he seemed to be paying to her. New issues such as compliance with dietary and exercise regimen were set as new goals for further joint cooperative efforts.

Case Vignette 26: Discussion

Mildred has been on a downward functional spiral due to multiple factors. She had undeniable medical problems of hip pain, obesity, and limited mobility, but her unrecognized, slowly progressive cognitive decline was further limiting her problem-solving ability and, in conjunction with the disinhibiting effects of the improperly consumed sedative medication, her ability to cope emotionally was also impacted. Her husband became the focus of her frustration, and he was simplistically blamed for all of her perceived unhappiness. Certainly, her husband's personality traits were contributing factors but this married couple had managed to raise three children and work out their differences for 38 years of marriage previously. The difference now was that Mildred, due to her cognitive impairment, was less able to be flexible, to see things from another's perspective, and showed poor judgment in overmedicating herself. With her husband educated about the "big picture" of her cognitive status, he began to view her as someone who needed him more than he thought she did. He seemed to rally to the cause and vowed to supervise her medications and take a more active role in encouraging her to comply with good health habits. Mildred perceived his efforts as renewed interest in her by her husband. The IPT-ci interventions seemed to allow for a new level of relatedness that was commensurate with Mildred's current cognitive abilities. Mildred's husband was relieved to "have things explained" and said he felt like he had a renewed purpose in "looking out for her."

10

The Caregiver's Own Role Transition

The first step for the IPT-ci therapist toward helping caregivers with their own role transition is to listen empathically. If it becomes evident that the caregiver cannot talk freely in the presence of the patient, then arrangements should be made to talk in private. Techniques to carry this out without generating undue suspicion on the part of the patient is to normalize the procedure as a routine part of the evaluation process, to ask a nurse or a support person to have the identified patient's blood pressure repeated for a "double check," and then to meet up in the waiting room afterwards. Alternatively, you can ask a coworker to conduct a parallel interview about health habits, or suggest that further information be exchanged at a later date by telephone, as you hand the caregiver your professional calling card.

With a background in the principles of gerontology/geriatric medicine and with added experience from working with other patients and caregivers, the IPT-ci clinician has hopefully developed an understanding and perspective about the common themes and the variety of ways that families have come to terms with them. For caregivers who have extreme difficulties coping with their role transition or who seem to be struggling with multiple problems in addition to caregiving for their cognitively impaired family member, it may be appropriate to diplomatically suggest that their own individual treatment might help them to cope better. If this suggestion is carried out and is successful, then the identified patient benefits indirectly from their caregivers' improved coping ability.

As illustrated in case vignette 26 in the previous chapter when a cognitively impaired woman was abusing sedative medication, helping caregivers to understand what they are seeing is key to forging workable coping strategies. Increased caregiving responsibilities are often experienced as a role transition for the caregiver and depressed mood is common among caregivers (Lu et al., 2007). Table 10.1 lists the various factors that affect caregiver stress.

Lending an empathic ear for caregivers who are struggling with their own role transition around the need to accommodate the cognitive changes seen in their loved one can shift their frame of reference. Adequate psychoeducation can improve caregivers' tolerance for frustration and enlist their help as proactive members of the support team for their loved one.

Table 10.1 Factors That Affect Caregiver Stress

- Coping with their own role transition to the stress of caregiver status
- Anger or frustration toward the identified patient
- Intrinsic coping ability (how well they usually handle stress)
- Other concomitant stresses—childcare, overburden, lack of help or support, and so on
- An historically difficult relationship with the identified patient
- Substance use or abuse
- Physical or mental health issues

Role Transition for Adult Child Caregivers

The role transition in adult child caregivers often revolves around issues of control that may be foreign to them. Some have anticipated this need and have gradually been making more of a presence of themselves in everyday problem solving such that there appears to be a seamless ramp-up of effort over time as the needs of their parent(s) increase. If the cognitively impaired parent lives alone, more comprehensive surveillance and intervention is required. For other adult children, the transition to caregiver is more difficult. For parents who were strong personalities or were perceived to be domineering, the decision to try to "tell them what to do" requires a distinct reversal in approach that was unthinkable for adult children up until this point. Faced with the alternative of allowing their parent to be at risk for such things as safety threats, social embarrassment, poor health practices, or financial ruin if their parent's behavior continues unabated, they usually conclude reluctantly that they must step up and take responsibility even if it means confronting the daunting task of insisting on a course of action against their parent's will.

Role Transition for Spouse Caregivers

If the caregiver is a spouse, the role transition commonly involves needing to take on tasks that the patient used to be responsible for because the identified patient is no longer able to do so. This often involves needing to take over driving duties, bill paying and other financial matters, cooking, shopping, laundry, cleaning, and sometimes coordinating yard work and repairs. For profoundly impaired individuals, if personal hygiene help is required and the caregiver is outweighed literally by his or her spouse, then cajoling them into bathing or dressing may be physically demanding. In the later stages of cognitive decline, if there is any oppositional behavior or disinhibition that carries a threat of physical violence, any physical strength disparity can be very frightening.

Some male caregivers who grew up in the current cohort of traditionally defined roles see their life's purpose as the breadwinner and their wife's role to be in charge of all domestic matters. When these men become spousal caregivers, the necessity of taking over domestic duties is a frequently cited role transition issue (or complaint). Sometimes their wives had managed the finances of the household too and this is an added change for them. The need to learn to cook or do the washing is not a matter of physical strength for them as the role transition issues are sometimes for women who try to assume traditionally male-oriented tasks such as yard work or car maintenance. Support from adult children, hired domestic help, and eating more restaurant meals are all ways of attempted coping.

The next case vignette (27) required an intense degree of joint and individual sessions to cope with a marital role dispute, which was also a profound role transition for the wife of the couple (the caregiver). A spousal role transition can also occur from the evolving role changes from a full give-and-take mutual partnership to a nonreciprocal relationship in which the caregiver must assume a surveillant role that feels more like a teacher monitoring a student for appropriate behavior or worse, like a mother who must take over every aspect of caregiving, such as encouraging good hygiene, safety, nutrition, and constant reassurance for the patient with very poor short-term recall. The following case vignette will illustrate such a struggle.

CASE VIGNETTE 27: EMBARRASSING RESTAURANT BEHAVIOR LEADS TO ROLE DISPUTE

A retired college professor we will call Roger with early cognitive changes is noted by his wife Rose to be forgetting the things she tells him a day or two earlier and both come in very upset after a restaurant debacle. While visiting one of their children in another city, they agree to take an entire group of family and friends out to a midday meal to celebrate a birthday. They are taken to the favorite restaurant of their adult child where they are friendly with the waitstaff from frequent visits. The surroundings are rather noisy and there are 12 in their party. When the meal has finished and, even though leaving has been mentioned, the party members are lingering in various animated conversations around the table. Roger is getting nervous and agitated that the waiter has not brought the check yet, which he has agreed to pay. He can see the waiter busily doing something in the distance and begins to conclude that he is being avoided for reasons he cannot imagine. Roger finally gets the waiter's attention loud enough for all to hear at the table and angrily calls him lazy and disrespectful to the shock of all present who themselves saw nothing to be annoyed about. His daughter was upset as she and her husband considered this waiter to be one of their favorites and were on very friendly terms with him.

CASE VIGNETTE 27 (*Continued*)

Rose was aghast with this uncharacteristic behavior and eventually the patient himself became very self-conscious, embarrassed, and ashamed—not at what he had said *per se* but by the collective distress he seemed to have caused, which he reasoned must have meant he had done something out of line even though it seem perfectly justified to him at the time.

This example is typical of the effects on social interactions of cognitive impairment characterized by executive dysfunction further disinhibited by alcohol. Several contributing factors helped to set the stage: (1) mild hearing loss made discriminating conversations difficult; (2) multiple conversations meant multitasking and rapid shifting between points of attention that eventually began to overload his brain circuit capacity to shift and follow conversations without becoming exhausted; (3) he also admitted to some anxiety about the size of the bill and was somewhat rigidly fixated on his need to "make sure he took care of it"; (4) room noise and other bustling activities in the restaurant added to the noisy confusion; and finally (5) there is the disinhibiting effect of wine he consumed with his meal.

In a joint session a few days later, Roger admitted that he felt terrible for causing such a scene that embarrassed the whole family and he vowed to do anything to avoid a repeat of the same. The action taken by the IPT-ci therapist in methodically explaining the above factors in the context of someone with cognitive decline was a huge relief for his wife to hear. Despite her college education, these "angles of observation," as she put them, were foreign at first but very familiar when they were clarified with other examples reiterated from what had already been described. Subsequent visits with Rose alone allowed her the opportunity to vent her frustrations and also provide additional history about a lifelong pattern of obsequious behavior on Roger's part until he would "blow his stack" with an angry vituperative outburst over something trivial at the moment. This pattern began in his childhood while living with his mother who suffered from bipolar illness and with whom "you never knew who you would meet coming in the door from school on any given day." In Rose's view, this early life experience left Roger with a quiet, contemplative disposition that was seen by others as "always calm and collected." Rose told of one of their infrequent visits to his mother, however, when Roger fell into a rage of screaming at her after he perceived that she placed one too many demands on him, however trivial they seemed to Rose at the time.

Roger's IPT-ci therapist also began to see that this lifelong pattern of behavior was also ameliorated by Rose who was attracted to his kind disposition and his intellectual skill, but it was also noted that she ended up taking over most social events and directing his interactions with others as he seemed poorly equipped to do so throughout their marriage. Roger did not seem to mind her subsuming the social calendar historically, as he had little interest in it on his own and was content to follow her lead. Both of them seemed content with their respective roles until recently. Roger's traits, in Rose's view, became exaggerated in recent months and were now "over the top."

In an individual session, Rose poignantly described her own sense of loss in that her husband no longer seemed able to be a full partner with her and that she now felt she needed to watch over him and "hover" at times to see to his needs and to help to avoid further social embarrassment when they went socializing as a couple. Rose acknowledged that she was going through a role transition too. She recalled how her own father had deteriorated from dementia, and she admitted to being "horrified" at the thought that her fate might be to care for a demented husband as well.

In that their marriage seemed to work quite well historically, it seemed logical and plausible that they try a joint coping strategy. They agreed to try working as a team in social situations such that she would diplomatically signal him when he was becoming inappropriate or embarrassing. Both agreed to give it a try and they left the session feeling more hopeful. In reviewing their first attempts at subsequent meetings, however, Roger admitted he did not like being told what to do and often reacted with rage, perhaps a throwback to his formative years with an unpredictable and demanding mother, he thought. As the alternative was pointed out to be a risk of more social embarrassment which neither wanted to see happen again, Roger agreed to discuss working out a subtle communication system, where his wife would ask him to "take a brief walk with her" during which time she would offer to help clarify anything confusing or to offer him a chance to problem solve a given situation with her help. Their struggle is ongoing at this writing with joint role transitions: Roger is struggling to improve his awareness of the effect he has on others but tension remains over his resentment of being "told what to do." Rose still struggles with resenting the fact that she has lost a spontaneous partner and does not want to be a "mother or teacher" for her husband. Both agree, however, that the outbursts and *faux* pas that brought them for help are things neither of them can continue to ignore. Rose eventually sought treatment for her own depression.

In the following case vignette, the patient requesting help for her depression turned out to be a caregiver for her dementing husband.

CASE VIGNETTE 28: STRUGGLING CAREGIVER COMES ALONE FOR HELP

Agnes, age 71, had been a participant in one of our clinical research studies for depression during which she showed a good response to antidepressant medication and was now in the long-term follow-up phase to monitor her

CASE VIGNETTE 28 (*Continued*)

progress. She had identified the stress of caregiving for her husband with Alzheimer's disease (AD) at home during her acute treatment but was now saying that she was feeling even more frustrated and stuck and did not know what to do to cope with her situation any longer. She was not currently receiving any psychotherapy, just follow-up medication review meetings. She asked if she could participate in IPT-ci to examine the ways in which the interface of her perceived caregiving duties and her husband's true needs might be reevaluated. In this case, the caregiver is the identified patient, although the IPT-ci principles apply in understanding her role in caregiving for her husband with AD. She did eventually follow through on the invitation to bring her husband with her to one session in order to complete the big picture and allow the IPT-ci therapist to meet him.

Agnes described her husband as a highly intelligent and prosperous businessman who was diagnosed with AD 4 years earlier and for whom she needed to take over more and more tasks. He was also 15 years her senior. Agnes described several role reversals in her role transition to full-time caregiver as he was formerly the social calendar keeper and his dynamism and energy had kept her happily busy during their marriage with extensive worldwide travel. She was now growing uncomfortable leaving him alone as he often forgot instructions and became fearful if she were not in the immediate vicinity to reassure him or to provide for his need of the moment. Agnes described her husband as conversational and he performed all of his own ADLs but was unable to help with any household chores and seemed content to watch TV and to read the paper although his poor recall precluded any chance of meaningful discussion of what he had just read. He had also occasionally made insensitive comments in public and became impatient easily and would abruptly ask to go home during outings, which discouraged Agnes from going places with him. She described having a "sinking feeling" that her life had been reduced to a locked-in, simplified routine that best suited her husband's needs but not her own.

When asked what she would like to do differently if she could do whatever she pleased, she mentioned reestablishing a prayer group which she had stopped due to the inconvenience of finding someone to stay with her husband and also that it was too disruptive to have it in her own home which she had done for years previously. She also missed going to plays, cultural events, leisurely shopping trips, and even getting her hair done as regularly as she used to. When asked about her interpersonal inventory, she described her friends within her church and her four stepdaughters (from his previous marriage) who lived in the same area but "were busy with their own lives." When asked if they visited regularly, she replied that they did but that the visits often required more effort not less for her since she prepared meals while they visited with their father who found the visits stimulating and enjoyable. They would occasionally go out to eat for lunch, but, this too required more effort to get her husband ready. On closer examination, it seemed that Agnes was frustrated because her stepdaughters did not seem to appreciate how impaired their father really was particularly since he remained quite articulate. Agnes was not sure how to ask them for more help. After several sessions of working on possible alternate approaches, Agnes concluded that what she really needed was to have visitors who freed her to do some other things for herself. A day program to occupy her husband and to provide safety and surveillance for him was discussed, but she felt he would not accept it. The therapist suggested that Agnes ask his daughters to take their father alone to lunch or stay and make him lunch while Agnes did other errands out of the home. She had to work up the courage to ask them but eventually did so and they agreed. Another option was to hire a companion to come and be with her husband on a regular schedule so that she could plan and look forward to a regular block of time for herself. She had never entertained this notion but found the idea doable and affordable.

After three individual sessions, she agreed to bring her husband along so that the therapist could meet him and make any other useful observations and he agreed to come wholeheartedly. The therapist informed him that he routinely met with the spouses of all his patients to ask them their opinion of what were stressors in their lives and how they were coping with them. Agnes's husband was as Agnes had described him: obviously intelligent, engaging, and highly articulate despite having very poor recall and little insight into the nature of the distress his wife endured in providing constant care and surveillance for him. He saw himself as retired with few worries, a stable marriage, a comfortable home, and children to be proud of. He had limited insight but did not seem to mind the idea of his wife going out more often on her own.

During a follow-up visit alone, Agnes reported that she had made some progress planning to inform her stepdaughters of their father's true level of cognitive decline that was not obvious on their brief visits and the fact that she could use more help from them. She decided it would be easiest to write a letter to each of them to be able to finish her argument without coming to tears. She succeeded in this effort, and her stepdaughters responded with statements to the effect "why didn't you tell us sooner." They said they would make more time to have "one-on-one quality time" with their father and give Agnes simultaneous free time to do whatever she wanted. They also agreed to help find a suitable paid companion for extra time during the week so Agnes could regularly attend her prayer group.

Agnes was thankful for the clarification she achieved during her IPT-ci sessions and the help and encouragement to get moving on a plan to change things for the better. Agnes also stated that she now realized that her husband had been such a dynamo in his prime that she had taken a "back seat" allowing him to set the agenda and had been more than content to have him

CASE VIGNETTE 28 (Continued)

orchestrate all of the trips and the busy social calendar they had once had. Agnes realized that she now had no choice but to accept the reality of his dementia at this time in their life but she now felt that she also had to admit to her own lifelong difficulty in asking for what she needed in many different scenarios and she decided that she wanted to reengage with a psychotherapist that she had seen previously to "do some more work on this issue."

Her IPT-ci therapist agreed to meet with her again in the future to do more problem-solving regarding her role in the care of her husband if she wanted to.

Case Vignette 28: Discussion

This case illustrates the flexibility of the use of IPT-ci for this caregiver who was feeling stuck and could not seem to cope effectively with the caregiving role for her husband with dementia. This case could have been handled effectively by a number of other approaches such as supportive psychotherapy, problems-solving therapy (PST), cognitive-behavioral therapy (CBT), or insight-oriented psychotherapy. As IPT-ci particularly focuses on the interface between the person with cognitive impairment and their caregivers, the initial presentation of the problem could also be approached from the cognitively impaired person's perspective although he was voicing no complaints. The caregiver in this case is the one struggling to balance the needs of her cognitively impaired husband with her own needs.

Considering the Role of Siblings

Among the varieties of possible family experience, siblings sometimes find themselves living in proximity to each other as they age. A mutual interdependence often develops, particularly as their parents' generation passes away. With the demise of spouses, siblings sometimes choose to cohabitate to share resources and to rely on each other in times of need such as during rehabilitation from an illness or surgery. If children or grandchildren are not in the picture for whatever reason, siblings will often say that all they have left is each other. If one sibling, not necessarily the older one, begins to show signs of cognitive decline, the other sibling sometimes becomes a caregiver. Unless there is a large age gap between siblings, both siblings may be suffering from some stage of cognitive decline and the distinction between patient and caregiver maybe less clear—both may need individualized assistance. Also, as with spousal caregivers, the siblings own physical disabilities or medical problems can limit their caregiving capacity (such as driving ability).

Sibling rivalry can still be evident at any age, although it usually dissipates with time. It may, however, limit a sibling caregiver's ability to cajole or encourage the afflicted sibling to accept the most reasonable course of action. In the current cohort of elders, who often grew up in an era of gender differences where the men usually worked outside the home, did most of the driving, and usually had the "final say" in decisions that needed to be made, these persisting gender lines can also spawn role disputes if the cognitive decline in one sibling requires a role reversal in the other.

The IPT-ci therapist tries to utilize all resources available to assist the needs of the patient to remain as autonomous as possible while simultaneously maintaining constant surveillance regarding the suitability of the caregiver as the therapist would with a spousal or adult child caregiver.

Friends Who Want to be Helpful

Friends can indeed be "true friends" when they help to maximize the quality of life for those suffering from cognitive decline. The term "friend" needs to be broadly interpreted. In addition to long-term friendships, we can also include lovers, neighbors, in-laws, partners of grandchildren, and so on. Some caregivers with no family ties have been champions of good care way beyond the call of duty. One such caregiver comes to mind.

CASE VIGNETTE 29: DAY-TO-DAY COPING HINGES ON FRIEND'S INTERVENTION

Jane befriended her neighbor Phyllis, who was 20 years her senior, and grew increasingly worried about her as the years wore on. The author got to know Jane as she dutifully accompanied Phyllis to all of her follow-up appointments for chronic depression and cognitive decline at our multidisciplinary geriatric clinic. Without Jane present, it did not take Phyllis very long to become disorganized, to forget to eat or to pay her bills correctly. Despite needing to care for her own family, Jane made time to visit Phyllis daily, to bring her home-cooked meals, to help her to pay bills, and to take her to all of her doctor visits.

The family history of Phyllis was a sad and unfortunate one. Her husband divorced her 15 years earlier leaving her a financial nest egg that should have guaranteed her comfort lifelong. Unfortunately, her two sons manipulated her out of most of her funds to fuel their drug habits, leaving her only with

---

CASE VIGNETTE 29 (*Continued*)

her home, for which she could barely pay the taxes and subsist on her remaining income from social security. Jane described being appalled by these events, but felt powerless to right the wrongs even though she tried by collecting all the documents and paper trails she could manage. In the end, there were no funds for lawyers and the increasing care demands of Phyllis made her drop the effort. As Phyllis continued to show worsening cognitive decline, Jane did her best to patch together a network of services such as meals on wheels and medication subsidies to stretch Phyllis's greatly reduced income and to keep her from losing her home. At the end of each month, when Phyllis was low on funds, Jane had more than once pitched in with her own money to pay utility bills. As Phyllis continued to decline into a clear state of dementia, Jane grew worried that she could no longer sustain these multiple efforts and felt that Phyllis needed assisted living to protect her and provide for her on a 24-7 basis. She did her research and found a likeable place but Jane's income was incrementally too high to meet the HUD requirement for entrance and she was becoming desperate. Jane's selfless pursuit of help for Phyllis was infectious and it rallied the author to call a local congressman to see if any other help could be found. Over the ensuing 18 months, Jane developed her own medical problems but rather than give up on Phyllis, she instead moved her into her own home.

At this point, Phyllis looked more robust than ever. She was well fed, always well groomed and her tendency to have tearful episodes where she complained of fears she could not articulate were quickly comforted by the numerous distractions Jane constantly provided. In effect, Jane and her family had made Phyllis part of their family. Without their magnanimous efforts, it is very clear that Phyllis would have soon required placement in a nursing home or a forced guardianship.

As this vignette clearly describes a case of elder abuse based on alleged financial manipulation by her own sons, it would be prudent to check into local statutes which might offer free or low fee legal consultation. Such a suggestion was made to Jane and Phyllis as well as a referral to the Area Agency on Aging to assess her eligibility for government subsidies.

Paramour Caregivers

Another scenario the author has witnessed several times involves a romantic relationship that develops between a widow and a widower, neither of whom wish to remarry but want to maintain an ongoing romance and close relationship. They may both feel that they are content to remain in their respective abodes and meet for "quality time" because they both have many other obligations to their respective children and friends they have accumulated over a lifetime and do not have the energy to try to create a blended schedule to try to accommodate what are often conflicting obligations. In the case where one member of the dyad begins to show cognitive decline, depression, or both, over time, the relationship can evolve into a caregiving one, especially if no adult children live near by.

Should this caregiving relationship become more burdensome over time, decisions need to be confronted about finances and other matters for which the paramour has no real control. Sometimes, role disputes arise between such caregivers and adult children of the afflicted partner on decisions that involve medical treatment, moving, driving cessation, or financial matters. The latter can bring fears in the paramour caregiver of being accused of financial misuse or in the most selfish context, if the paramour is viewed as being less well off financially, fears on the part of adult children that inheritances might be usurped. In some instances, however, the adult children's suspicions are correct.

The IPT-ci therapist is sometimes put in a position of not knowing who to believe as the interfamilial forces surrounding the control of money can be quite strong. It is not the responsibility of the IPT-ci therapist to be a detective *per se* but rather to try to keep an open mind and to try to ascertain where the friction lies between all parties who claim to have a stake in the welfare of the identified patient. It is at these times that IPT-ci therapists should remind themselves that they are an advocate for the identified patient and that if it appears that actions are being taken that are at cross-purposes with this mission, then encouraging necessary steps to rectify the problem is indicated, including referrals for legal or protective services.

11
Flexible Individual and Joint Sessions

In contrast to traditional IPT, which was designed as an individual treatment with an allowance to meet a significant other at some point for clarification of their respective social role(s), the core of IPT-ci is its comprehensive input from the patient and all involved parties whose role may be evolving into caregiver status from the first meeting. When IPT-ci therapists meet a new patient, they are prepared to be flexible according to the unfolding needs of the identified patient as the particular situation calls for as outlined in the Table 11.1.

As you can see in Table 11.1, for patients who present alone and who appear to be cognitively intact and seeking help for depression, they may be best served by traditional IPT in which family members are involved either at the patient's request or at the therapist's suggestion that it would be beneficial to meet a significant other to better understand the whole picture. When patients come with family members who are sitting with them in the waiting room, you can call the patient's name and see who stands up. If they all do, then start your evaluation with all parties. If there is hesitation on the family's part about whether to go with the patient, ask the patient what he or she would prefer and tell him or her that you are comfortable either way and that the decision is up to him or her. For patients with early cognitive impairment, they may feel unsure of themselves or nervous and are comforted by having their family come in with them. Sometimes, a determined family member will stand up as if to say: "I want to be sure you know the whole story, so I better

Table 11.1 IPT-ci Engagement Variations

Who Comes?	Apparent Diagnoses	Strategy	Engage Family?
Pt. alone	Depression only	Trad. IPT	Patient preference
Pt. & fam.	Depression only	Trad. IPT	Patient preference
Pt. alone	Cognitive imp.	IPT-ci	Strongly encouraged
Pt. & fam.	Cognitive imp.	IPT-ci	Strongly encouraged
Pt. alone	Dep & cog. imp.	IPT-ci	Strongly encouraged
Pt. & fam.	Dep. & cog. imp.	IPT-ci	Strongly encouraged

come in there." The patient might shrug his or her shoulders in ambivalent acquiescence.

Early Visits

The psychoeducational component of IPT-ci seeks to educate all parties, particularly if the diagnostic picture becomes more clear over time to reveal significant cognitive impairment. It is important to remember that you are beginning an evaluative process by establishing good rapport and a good flow of accurate information (from all available sources). It is important to explain to all present that the way you work is to spend time interviewing each concerned person (separately if indicated) to get the most accurate picture of who everyone is and how they fit into the life of the identified patient. Tell them that you also would like permission to gather old records, and labs or other testing reports. When a clear picture emerges of what help would be best for the identified patient, inform everyone that you are happy to meet with the entire group to explain your conclusions and your proposed plan of action for their review and comment. In reality, there is usually a reason given for patients coming such as a referral from their primary care doctor, emergency room, or a record of the initial telephone call made by a family member outlining their chief concern. This information, if available, will allow you to prepare a strategy for how you plan to engage the patient and family.

The most important thing you want to achieve at the first meeting is to make all parties comfortable with you as an advocate for the best interests of the identified patient and establish connections with all other concerned parties who will help to inform the ongoing treatment plan. Seeking explicit permission from the identified patient at this time allows continuous input from those accompanying parties and sends a message of respect for the autonomy of the identified patient. Leaving all parties with your phone number is another way of making it clear that you expect to hear from any or all of them. To be absolutely legal from a HIPAA standpoint or privileged healthcare information standpoint, you can ask the patient to list all the names of concerned parties on the permission-to-contact form such that they are authorized to call and exchange information without worrying about confidentiality issues.

Once the conduits of information are in place and the interpersonal inventory has been pieced together from all available sources, the IPT-ci therapist can begin to understand the particularly needs of the patient, any looming decision points or impending crises, the extent to which meaningful communication is taking place between concerned parties, the degree of psychoeducation required, and finally, any role disputes that are complicating the picture. Once this information is obtained, the IPT-ci therapist begins to formulate a strategy for maximizing the quality of life for the identified patient. This mission should be stated explicitly. If depression is the primary problem, then all the tenets of traditional IPT are put into effect, which include explaining the plan to the significant other or accompanying companion. Antidepressant medication may be indicated or even electroconvulsive therapy (ECT) may be conducted for severe cases. These somatic therapies are perfectly compatible with IPT and IPT-ci and a review of compliance with medication is a crucial component of the monitoring process. If cognitive impairment is part of the picture, a work up or evaluation of possible causes is warranted in collaboration with the patient's primary care physician or specialty consultant. If the problem is diagnosed as dementia, a trial of one or more memory-enhancement medications may be in order.

After rapport is established, complete information from all sources has been gathered, and dementia work-ups and any indicated medications are begun, the IPT-ci therapist then begins in earnest to work on the agreed-upon focus of either grief, role transition, role dispute, or interpersonal deficit/sensitivity. If the identified patient can come as regularly as weekly, that would be ideal for making therapeutic progress. If weekly visits are not possible, then attempting to meet as regularly as possible is striven for. Take what you can get. Do your best to "move forward" therapeutically every chance you get. The telephone can be used to bridge lengthier time gaps between sessions and this includes telephone contact with the identified patient as well as other pertinent family member or caregivers as necessary. Hopefully, the IPT-ci therapist is perceived by all parties to be a dedicated professional who takes time to listen to everyone's concerns and seems genuinely interested in the "big picture," which includes the opinion of all parties and the welfare of the caregivers as well as the identified patient. These early visits are also crucial opportunities to monitor rigorous compliance with psychotropic drug regimens and to relay feedback about potential side effects to the prescribing physician for adjustment decisions.

The timeframe for follow-up visits from this point onward may be variable and at the discretion of the IPT-ci therapist. If the patient is cognitively intact and is being treated for depression with traditional IPT, then weekly attendance and forming the usual contract for 12–16 acute sessions is strongly encouraged. If a cognitively intact older person were to successfully complete the course of IPT with full resolution of depressive symptoms, then termination as planned might be appropriate. Those patients with a history of recurrent depression, especially if the IPT problem area was role dispute, might ideally benefit from monthly maintenance IPT sessions for long-term prophylaxis against a recurrence of major depression.

How to Talk With Patients and Caregivers About Dementia

If the diagnostic work-up for patients showing significant cognitive impairment indicates early Alzheimer's disease or other potentially progressive

Table 11.2 Talking Points for Softening the Impact of a Dementia Diagnosis

1. Having a diagnosis of AD does not mean being one step from needing a nursing home.
2. AD is usually slowly progressive over a period of years or even decades.
3. It doesn't mean you have to stop the things you enjoy, on the contrary, staying busy and keeping your mind stimulated may help to fight further memory loss.
4. You are not alone with this diagnosis. The Alzheimer's Disease Association and other organizations have a lot to offer to help you and your family to learn more and cope better with this diagnosis.
5. Medications can help. Aricept®, Exelon®, and Razadyne® are so-called cholinergic enhancers that have shown an ability to slow the progression of AD. Exelon® is available in a once-a-day patch as well as a pill. Namenda® further helps to slow the disease process and is usually used in concert with one of the above as it works through a different mechanism.
6. Staying well informed with the latest new developments in AD research can give a sense of hope that breakthrough discoveries may be coming.

dementia, then planned termination as in traditional IPT does not make sense. These patients, unfortunately, may be on a downward course of unpredictable speed. The best somatic treatments for Alzheimer's disease at this time do not arrest the disease process although they may slow it down. Victims of progressive dementing disorders can sometimes "plateau" or even improve somewhat with adequate treatment of any comorbid depression and with the introduction of memory-enhancing medications or other therapies, but their long-term prognosis is still relentlessly downhill. This unhappy news is not something to accentuate but it is the responsibility of the IPT-ci therapist to honestly answer any questions put to them by either the identified patient or other concerned family members without sugar coating but maintaining a hopeful tone that ongoing research may yield breakthroughs.

Table 11.2 lists some "talking points that might prove useful to "soften the blow" of a diagnosis of AD or other progressive dementia.

Long-Term Follow-Up

Utilizing IPT-ci for long-term follow-up and advocacy for patients with dementing disorders makes the most sense in a similar way that maintenance IPT is used for long-term depression follow-up. The follow-up schedule for IPT-ci is flexible; monthly follow-up sessions may be appropriate or if the patient has reached a plateau of reasonably good functioning given their cognitive impairment level, follow-up visits every 3–6 months may be warranted. A "steady state" is achieved when the patient and all concerned family have been educated about depression and cognitive impairment including dementia as well as educating them about how to recognize signs of worsening depression or further evidence of cognitive decline for which they can call and make an earlier appointment for reevaluation between scheduled follow-up visits. Further evidence of cognitive decline that results in new safety concerns, role disputes, or a need for a higher level of care (such as a move from a private home to a long-term care facility) or, the added stress of an intercurrent medical illness or surgery, would indicate the need for a period of closer follow-up. The goal of follow-up would then be to reestablish a new steady state of functioning from a safety and well-being point of view regarding the identified patient's cognitive status and mood, as well as a reassessment of the current coping ability of any involved caregivers.

12

Reaching Steady State and Long-Term Planning

Relieving depressive symptoms has always been the goal of IPT, and maintenance IPT has been shown in one study of geriatric-aged patients to have a protective effect against a recurrence of depression, particularly in those whose IPT focus was role dispute (MTLD-1; Reynolds et al., 1999). As outlined in the previous chapter, a diagnosis of cognitive impairment or dementia with a long-term prognosis that portends further declines in cognitive function requires caregivers to be involved in making decisions in the best interest of the identified patient. The modifications of IPT-ci outlined in this manual are intended to help the IPT-ci therapist to engage and educate both patient and caregivers, to be an integral part of a team effort to evaluate possible causes for the cognitive impairment, and to institute appropriate treatment that includes psychotropic medication as well as IPT-ci that provides psychotherapeutic benefits for the identified patient and the caregivers. The range of interventions in IPT-ci consists of clarification, continuous psychoeducation, problem solving, role dispute resolution, communication analysis, decision analysis, the testing of perceptions and performance, and sometimes role playing. The IPT-ci therapist also tries to remain sensitive to any role transition the caregivers may be going through, assesses their suitability for effective caregiving, and offers advice for referral for extra services to assist them in caregiving or to get help for themselves as indicated.

Steady State Maintenance

Steady state is defined as the point in time when the IPT-ci tasks have been achieved, the patient's depressive symptoms have been minimized, and whatever cognitive impairment that is present has been elucidated and treated, resulting in optimized functioning with appropriate accommodations put into place to provide supervision, assistance to guarantee safety, and optimal stimulation and socialization commensurate with their degree of cognitive impairment.

Once a steady state is reached, it very important to continue maintenance visits long term and to be prepared to reevaluate the status quo. Some new event such as a medical complication or cognitive function that appears to be worsening may require one or more changes to be made for safety or increased accommodation for daily functioning. Such changes may or may not be readily agreed to by the identified patient and another role dispute may erupt in which the IPT-ci therapist can prove helpful to broker a new agreement that is the most reasonable solution or compromise. Having no prior experience, caregivers may find each new level of cognitive decline to be a new challenge to cope with and require the input of the IPT-ci therapist to cope optimally. For family members or caregivers who have been thoroughly educated and have a strong grasp of the reality of the big picture, problem solving and adjusting to new contingencies might be handled without consultation of the IPT-ci therapist. This would indicate that the IPT-ci therapist had done a good job of preparing the caregivers to anticipate and handle the situation on their own.

Caregivers and patients might well drop out of treatment for a variety of reasons (moving, nursing home placement, etc.) but regular maintenance visits of an appropriate frequency, as judged by the IPT-ci therapist, can serve to validate decisions, foster a spirit of hopefulness rather than demoralization, and continuously reassess the caregiver's stamina, mental health, and resourcefulness to cope with any noticeable changes in cognitive impairment in the identified patient over the long term.

Long-Term Planning

We have all read occasional accounts in the newspaper of older individuals who had deteriorated slowly from fully independent living to a point of severe cognitive impairment such that they were found to be living in squalor and in need of adult protective services to intervene and place them in a nursing home as they clearly could not take care of themselves any longer. This situation proves that not everyone has willing and available concerned family or caregivers to look out for their long-term interests. A forced move in such situations, although necessary, is often traumatic for elders who perceive the action as wrenching them from their home and suspending their right to self-governance as the following case vignette will illustrate.

CASE VIGNETTE 30: EMERGENCY PLACEMENT FOR A COGNITIVELY IMPAIRED WOMAN

As the psychiatrist member of a team collaborating with an Adult Services agency, I evaluated a woman in her home who was reported by neighbors to be behaving in an odd way. She had also been threatened with having her utilities shut off for nonpayment. She greeted me openly and invited me in but did not seem to care or question who I was or why I was there. In following her up a set of stairs into her home, I noticed what appeared to be a wadded mass of material in her slacks that I guessed to be rolled up tissue to cope with incontinence. Her home was piled high with two-foot stacks of unopened mail on every surface, and there were stacks of unfolded newspapers that were clearly never opened. The sink was piled high with dirty dishes such that there did not seem to be any clean ones left. There was mold on the dishes indicating that they had not been cleaned in a long time. She had a cat that had obviously urinated and defecated on the carpets repeatedly as there was evidence of feces and a strong stench of urine. She was quite thin and no doubt nutritionally compromised from poor intake. She was articulate but rambling in her speech and she could not tell me the day or date. She spent a long time in the bathroom which made me wonder about incontinence or just impairment in toileting ability due to cognitive impairment. This woman clearly had no supervision and was quite demented and functioning at a dangerously poor level such that I recommended to the Orphans Court that she be declared incompetent and taken into a treatment setting and eventually a nursing home for her own good. This procedure was carried out despite her protests and a guardian was appointed for her person to make future decisions in her best interests going forward.

Case Vignette 30: Discussion

In reviewing her old records, it turns out that a colleague of mine had evaluated her some 10 years earlier and had actually treated her for depression. When I contacted him, he recollected that he suspected some early cognitive impairment at that time, but she was employed as a scientist in a local university and with her high intelligence she was compensating quite well at that time. She was lost to follow-up and clearly had "fallen through the cracks" in a profound way.

This sad case of profound cognitive impairment that went unnoticed until it placed the person in dire circumstances obviously could have been prevented with regular follow-up care and appropriate planning, which could have resulted in a safe and dignified segue into more intensive services as her cognitive impairment worsened.

In this case, a guardian had to be appointed for her person to make future decisions in her best interests going forward. If no family members or friends are available or willing to perform this task, some other person must be sought out through appropriate legal channels. In Pittsburgh, for example, there is an order of nuns called the Ursuline Sisters whose spiritual mission is to take on such cases and be their guardians.

Planning for Future Contingencies

After a steady state is achieved in IPT-ci and the identified patient is functioning at a sustainable level for a period of months, this may be a good time to diplomatically introduce discussion about planning for future contingencies. Some patients spontaneously ask such questions as, "Am I going to end up in nursing home?" or "Who will look out for me if I cannot take care of myself and my money anymore?" Even if these questions are not brought up, it is appropriate for the IPT-ci therapist to raise them for consideration such that planning can begin in earnest to anticipate such needs and to avoid the need for emergency action such as the necessity for appointing a guardian as the prior case vignette illustrated.

IPT-ci therapists should familiarize themselves with the particular laws in their locale for such things as power of attorney, last will and testament, long-term care insurance, guardianship, and a working knowledge of other geriatric support services. Social workers are the best trained in these issues, and if the IPT-ci therapist is not a social worker by training, then some consultation may be required. If the identified patient's cognitive status has deteriorated to the point that he or she may soon require a move to long-term care facility, having looked at them ahead of time can diffuse the anxiety in the patient and the caregivers and help them to make the most reasonable choice. Most facilities have waiting lists, which can take a patient months if not a year or more to become eligible for entrance. Even if the identified patient is not currently near the point of needing a move to such a facility, having the discussion about what it would take to trigger such a move well ahead of time is still a valuable exercise. Such issues as costs, available assets, location, and the ambiance of the facility are all factors to consider. Some facilities that cater to more affluent clients will serve lunch to visitors and even allow prospective clients to stay overnight to test out the ambiance of the place.

CASE VIGNETTE 31: SIMPLIFIED LIVING ARRANGEMENT EASES ANXIETY IN COGNITIVELY IMPAIRED MAN

Craig, aged 82, was engaged in treatment for several years for generalized anxiety disorder and depression along with multiple medical problems including a diagnosis of sleep apnea. He was originally treated with psychotropic management and traditional IPT.

The IPT issue that achieved good resolution early in treatment was the improvement in the relationship with his sons whom he had been estranged from since his first wife had died of liver failure from alcoholism at a time when he was also an active alcoholic. He subsequently was able to remain sober and remarried his current wife. The earlier course of traditional IPT had helped him to work through and cope with all the resentments his sons held about their teenage years living in a home with two alcoholic parents, particularly, their perception that had he been sober he might have been able to do more for their ailing mother (they blamed him partially for her death). He now met and talked with both sons on a regular basis, which fulfilled his one remaining wish in life to get to know his grandchildren whom he had been estranged from previously.

Over a several year period of follow-up visits every 3 months, he began to show signs of cognitive impairment characterized by poor problem solving, short-term memory loss, quickness to anger with small annoyances, and becoming easily bewildered when presented with more than one task (poor multitasking). His neuropsychological testing showed MCI, which progressed over the following 2 years to a diagnosis of early probable Alzheimer's disease for which a memory-enhancing medication was added to his regimen of antidepressant and antianxiety maintenance medications. He lived with his cognitively intact wife in their home of 15 years. They jointly agreed that they were getting too old to care for their home and planned to move to a luxury apartment complex on the grounds of a comprehensive facility with assisted living and nursing home options as they were needed (a life-care community).

Anticipating the logistics of such a move, however, raised Craig's anxiety sky high, which in turn affected his ability to think clearly and make rational decisions. This led to increasingly frequent role disputes with his wife. Some of his psychotropic medications were adjusted to help him to contain his high anxiety and to assist in combating his insomnia, which always worsened when he was under stress. This new level of decompensation necessitated a new round of IPT, this time using IPT-ci principles that specifically involved his wife who was transitioning to caregiver status and was also highly frustrated. In listening carefully to the daily pattern of Craig's struggles, it became clear that he and his wife argued regularly in the evening about how things should be done the follow day to prepare for the move. As this sounded a lot like "sundowning" (they did not seem to argue at other times of the day), his IPT-ci therapist made the suggestion that he and his wife avoid any decision making debates after their evening meal but rather save them for the morning when they were both fresh and rested. Craig and his wife, who attended several joint sessions together, agreed to try this new approach as they both stated their worries that these arguments were having a harmful effect on their historically good marital relationship. They were willing to try anything to find a way to stop bickering. After carrying out some communication analysis of these new role disputes, it became clear that Craig's wife was taking all of his anxious worries at face value and trying to argue logically that there was no good reason for him to be so anxious. A separate meeting with her outlined the connection between Craig's intense anxiety and his cognitive impairment which limited his ability to problem solve, particularly in the evening when

CASE VIGNETTE 31 (*Continued*)

his brain was tired. By using specific examples, she was able to grasp the big picture better. Avoiding evening "decision making" worked surprisingly well to reduce their role disputes and allow them to prepare adequately for their move.

Now that they reside at the comprehensive care facility, they are both enjoying it and Craig's anxiety is minimized. They are taking advantage of on-site planned activities and cafeteria style meals they can choose to partake of while still maintaining their independence.

Case Vignette 31: Discussion

The simplification of bill paying to a once monthly fee with no utilities, homeowner's insurance, repairs or lawn care duties was a great relief to both Craig and his wife at this stage of their life and the move clearly reduced the tendency for Craig to become overwhelmed with decision making that overtaxed his current cognitive ability. The role of his IPT-ci therapist, in this case, was to help Craig and his wife understand the source of his increased anxiety, point out Craig's difficulty multitasking and problem solving that led to new role disputes and subsequently, to make concrete suggestions for handling the stress better by seeking ways to simplify the decision making that was now too complex for Craig to handle in large measures, particularly when fatigued at the end of the day.

The next vignette will illustrate how the IPT-ci therapist had to contend with a role dispute as well as role transitions in both the identified patient and his spouse in working toward a steady state.

CASE VIGNETTE 32: WANING COGNITIVE ABILITY COMPLICATES DEPRESSION MANAGEMENT AND LEADS TO ROLE DISPUTE

William, an 82-year-old retired healthcare professional, presented for treatment of depression. His wife was 10 years younger and was still working. William had had prior bouts of depression that were reasonably responsive to antidepressant medication. William came by himself in the early sessions for medication management. His demeanor was quiet and deferential, but he had a hard time describing what he did on most days. He seemed to be too eager to accept suggestions without questioning as if he couldn't think of much himself. He seemed to have a paucity of thought and very little initiative. He scored 27/30 on the MMSE missing the correct date, he made one error in serial sevens, and he could only recall 2/3 items. When asked about hobbies, he mentioned fine woodworking, but he said he was not working on anything currently. Two adult children lived in distant cities, and he spent most work days by himself while his wife worked but he was poorly able to describe how he spent his time.

For a retired professional, an MMSE score of 27/30 can represent a large decrement in cognitive ability. It was not clear at this point how much his cognition was slowed further by his depression. William showed lots of body language consistent with a depressed person: he slouched, often held his head low, spoke in a low volume voice, and often was noted to hold himself with his arms wrapped around its opposite as if trying to somehow comfort himself. An IPT-ci focus of role transition to a state of lower functionality was established along with readjusting his antidepressant medication. He had stopped maintaining any contact with other retired peers. When asked about the possibility of rekindling his woodworking hobby, he consented, but in subsequent sessions seemed to have little initiative to actually begin. He did, upon request, bring in photographs of previous completed projects, which were intricate and beautiful. It was beginning to become clear that he was no longer able to measure up to his own high standard of skill and precision or even to begin what were obviously a complex series of required steps to achieve a finished product.

William's wife accompanied him on some visits. She voiced frustration that he seemed to only do things around the house if she insisted and was overly willing to follow her lead without seeming to have any plans of his own. She said she "wasn't ready for this," referring to her constant need to remind him to complete household tasks. A process of psychoeducation was begun with her to point out that his lack of initiative and lack of recall and follow through were likely due to his cognitive impairment, made worse by another bout of depression.

In a joint session, we all discussed the fact that William seemed to be at a loss as to what to do with his time and that he was also not deriving any satisfaction from any accomplishments as he once did when he was working at his job or doing his woodworking. His wife pointed out that he had many unfinished projects, and she wished he could "get back into it." In exploring this possibility further, we discussed whether she might help him to select one such project and begin with something basic like doing some of the fine sanding, but this too seemed to be overwhelming for William.

As William seemed to show strong depressive features, subsequent sessions focused on adjustments to his medications using combination therapy as his depression was now falling into the treatment resistant category after various failed trials of antidepressant medications. Over the ensuing several months, William made a modicum of progress in improved mood. We discussed the possibility of ECT, but he refused to consider it. His MMSE fell

CASE VIGNETTE 32 (Continued)

to 24/30 within a year. William's wife decided to retire, partly due to her increasing worry about him, and she now accompanied him to all of his visits. She continued to voice frustration that William seemed overly dependent on her to initiate everything from meals to household tasks to social events. William also seemed to be very sensitive to his wife's criticism, and he stated that he felt he was "holding her back." His wife, being ten years younger, also had an independent agenda for her own retirement such as having lunch dates with her other retired friends from her former workplace and expressed her wish that William had some of his own activities that did not require her input.

As William's depression was beginning to improve, he was sent for neuropsychological testing, the results of which showed a pattern consistent with dementia of the Alzheimer's type. This diagnosis was discussed with both William and his wife. He did not seem very upset and did not voice any concerns about any implications. His wife said she had been preparing herself for such a diagnosis and stated that she guessed that explained a lot of his behavior. William was begun on the cholinergic-enhancer donepezil (Aricept®).

In a continued search for a meaningful activity to help bolster William's self-esteem, at his wife's suggestion, he began to attend a local model railroad society meeting. At first, William was quiet, passive, and expressed some discomfort in not knowing anyone at these meetings, but he gradually warmed to the other members who invited him to participate in making repairs to the displays. For a retired professional known to be good with his hands and a woodworking enthusiast, this venue seemed well suited to him, and he began to attend several days per week for a few hours which allowed his wife to have unrestricted time for her own needs.

Over the ensuing year, William's memory began to deteriorate further, particularly his short-term recall, and his MMSE dropped to 22/30. Interestingly, William seemed to accept his memory loss better at this point and would volunteer that his memory was poor to new people he met. He seemed to be accepting his life as best he could under the circumstances.

Case Vignette 32: Discussion

In the past 20+ years of working with cognitively impaired individuals longitudinally, the author has noted similar progressions of cognitive impairment many times. In the early stages of cognitive impairment, the afflicted individual often describes a vague sense of tasks seeming to be "too difficult," which leads to demoralization as they are acutely aware that this was once not the case. Depression can worsen as an awareness of further cognitive decline precludes more of the person's former activities and, together with the lack of initiative in William's case, the afflicted person feels a sense of confusion, and a lack of focus or direction. At these times, having someone to organize daily life activities gives such individuals some direction to follow, although they may not follow through with satisfactory completion of the task in the mind of the caregiver as they forget details or lose track of the "big picture" or face a problem along the way that has become too complex for them to overcome. There seems to be a point reached in the progression of the dementia, however, when there is better acceptance of the status quo such that mood can be seen to brighten as in the case of William (his antidepressant medications and the Aricept® may also have contributed). This noticeable change in acceptance of their fate (having memory loss) has been attributed to losing insight or to loss of the "third eye" (or observing ego in psychodynamic terminology). William and his wife seemed to have reached a steady state of acceptance of the reality of William's cognitive state and their individualized compensatory adjustment to it. This may, of course, change with further decline to a new level of disabled function in the future, which will require their IPT-ci therapist to help all concerned to reevaluate the true picture and explore the best compensatory response to the new challenge in a continuous cycle for the remainder of William's lifetime.

The Perils of Last Wills and Testaments With Family Rivalry

As has been mentioned earlier, the IPT-ci therapist remains an advocate for the identified patient and as such, circumstances can arise where it appears that the selfish interests of others are taking precedent. There are many accounts of elder abuse documented in newspapers and agencies that catalogue such occurrences. Financial abuse is one such scenario that IPT-ci therapists can become aware of by virtue of how well they have gotten to know the patient and the caregivers. When an older person has assets that will likely remain after his or her end-of-life care is paid for, there can be competing interests that come forward to claim their right to it. A properly executed last will and testament would ideally put to rest any ambiguity about who receives what share of the estate but, in reality, it does not always work out that way. Wills can be contested, particularly if a new one was drawn up to supplant a previous will in the immediate time period before the person dies. The debate then centers on whether the late person was competent to know what he or she was signing and if not, the latest will might be invalidated in a court of law. Those relatives with selfish or competitive intent have historically tried all manner of ways to secure what they consider to be their rightful due.

It is not the IPT-ci therapist's job to sort out legal issues such as these but, after witnessing various abuses that seemed to run counter to the patient's wishes, the author is inclined to pay more attention to these developments as they unfold. For example, if patients have achieved a steady state and they are open to a discussion of long-term issues, then asking whether they have a last will and testament they are comfortable with is a reasonable question. If they do not but show interest, you might suggest a short list of local lawyers who could help them secure one, or discounted legal services if they cannot afford the fees of a private lawyer. If patients are suffering from some level of cognitive impairment, that does not mean they are incompetent to create a valid and binding last will and testament. The question of competency is a legal concept that requires some judgment of the patients' ability to understand to whom they want their estate proportioned and that they have a good grasp of what their estate entails. They also need to be free from the influence of designing individuals such that it can be reasonably assured that they are creating the document of their own free will. If you are aware that they are going to create such a will, making sure to document your opinion of their ability at that time to meet the aforementioned requirements in their medical record or chart will shore up any arguments about verifying that patients were competent at the time, should it be challenged later. Some patients videotape their last will and testament such that some judgment can be made as to their appearing competent on video compared to merely seeing their signature on a new document that could have been coerced or manipulated by another party.

What many people don't realize is that one also has the legal right to keep their last will and testament secret until after death has occurred. The author has sometimes suggested this option to individuals who felt they were being pressured by rival family camps to be sure they were positioned for their rightful distribution. In one such case, the patient wanted to will the bulk of his estate to his nephew who had dutifully visited him on a regular basis and helped him in his later years in contrast to his own children, who seemed disinterested, comparatively. When this man was on the verge of being placed in a nursing home, his children all showed up and wanted to know the legal opinion about whether he was competent to create a will at that point. The author was consulted by the court to render an opinion. After doing the evaluation, the author concluded in his report to the judge that although the patient was fuzzy about which banks housed all of his monetary assets, he was clearly able to identify which proportion of his estate he wanted to will to particular individuals (namely, the majority to his nephew). The author further recommended that the court appoint a lawyer for the patient to carry out his right to keep his new will sequestered, only to be opened after his death as he did not want to face the family squabbles regarding fairness as it was his decision alone.

Caregivers have been known to show up to take care of an ailing person and then position themselves to manipulate the elder to create a new will that identifies them as the beneficiary. On the other hand, selfless caregivers who devoted large amounts of time, energy, and their own money to the care of their aging relative are often not compensated disproportionately in the will if it was made years earlier with an equal share for each child or next of kin, for example. The IPT-ci therapist is cautioned against becoming too entangled in family disagreements or legal confrontations along these lines; however, turning a blind eye to that which seems evident is also contrary to the patient's best interest. Getting legal advice from a risk management point of view in your institution is sometimes warranted if you are unsure how to proceed or if you suspect possible elder abuse or fraud.

Advanced Directives: We All Need Them

Finally, power of attorney and advanced directives are important topics for any adult to have documented. They are not usually the pressing issues that bring identified patients for evaluation early in the acute phase of IPT-ci but, when a patient reaches steady state, it is prudent and protective of the patient's wishes to have them in place, particularly if there is cognitive impairment developing that might progress to the point where competence to make such decisions could be questioned. For example, the author has commonly seen elderly individuals with cognitive impairment who enter a hospital for acute medical treatment that might require consent to be given for a diagnostic procedure or surgery. If the patient shows evidence of cognitive impairment, the medical team may be reluctant to proceed unless there is a duly executed power of attorney document for medical decision making that allows another individual to make the decision on the patient's behalf or to validate the patient's concurrence. Hopefully, such an agreement was created before the patient's cognitive impairment progressed to the point where the patient's competency to make such decisions was in question. If no power of attorney and no next of kin is available and the patient appears to be incompetent to decide for himself or herself (the legal test is whether the patient can reiterate the essence of what is being recommended back to the doctor in his or her own words), then the medical team must seek temporary guardianship in a court of law to assign the rights of the patient to another person to act in the patient's best interests. This is a cumbersome and time-consuming procedure.

Advanced directives record the preferences of the person while the person is competent to make them. These documents outline one's preferences about how one would like to be treated should one become unable to speak for oneself in time of severe illness. Such decisions include whether one wants cardiopulmonary resuscitation, intubation, feeding tubes, or no extraordinary intervention beyond offering nutrition and hydration that must be ingested by the patient's own volition. These directives are designed to preclude lengthy artificial maintenance of life without subjective quality to be predetermined by the person for whom these decisions will affect at a later point (when the patient becomes too ill to decide for himself or herself). It is within the purview of the IPT-ci therapist to help to explain these concepts and to help patients to cope with any emotional reactions they may have to them and furthermore, to help to steer patients to those who can properly document their preferences such that they are actually carried out according to their wishes.

Many hospital facilities have their own forms to document these preferences and often require that they are completed upon entering a hospital (or to have copies brought along if they are already completed).

To minimize the cost of legal fees to loved ones, preprinted advanced directives that can be easily customized are available. One such document entitled "Five Wishes" was created by a consortium of lawyers. This document is easily readable and can be filled out by the individual with a pen using checked boxes and write-in blanks. When appropriately witnessed, it is considered to be valid and binding in several states. This Five Wishes document is often distributed by hospice agencies and is also available through the Internet for a nominal fee (www.communityhospice.com/getinfomed/fw2005.pdf).

Postscript

Future Directions for IPT-ci

The preceding 12 chapters are intended to be a manual of guidelines for utilizing a modified version of IPT for older individuals with cognitive impairment and/or depression along with their caregivers in the primary care setting. Interpersonal psychotherapy (IPT), by virtue of its pragmatic and user-friendly format, is an ideal psychotherapy for use with depressed persons, including those of older age. The IPT-ci modifications further address the needs of cognitively impaired patients.

Summary of IPT-ci

The major changes that make up the IPT-ci modifications are the integration of caregivers into the entire therapeutic process, with the recognition that caregivers need a great deal of education too and recognition of the fact that they are often going through their own role transition. The IPT-ci interventions remain primarily focused on the needs and welfare of the identified patient, however. Additionally, IPT-ci utilizes flexible individual or joint sessions with either the identified patient, caregivers, or both as needed for problem solving or role dispute resolution. There is particular emphasis on finding ways to understand and to cope with the executive dysfunction component of cognitive impairment that is so frequently misunderstood or misattributed by caregivers. The goal of psychotherapeutic work in IPT-ci (with both the patient and the caregivers) is to successfully achieve a steady state of optimized functioning along with optimized biomedical evaluation and treatment of depression and cognitive impairment.

IPT-ci can then also serve as a template for maintenance or long-term follow-up in order to remain surveillant for recurrences of significant depression and also the likely scenario where cognitive impairment worsens over time. Long-term follow-up sessions can be of variable frequency as judged appropriate by the IPT-ci therapist to reevaluate any changes in function whether they are related to depression, further cognitive impairment, or their combination. In the follow-up period, the IPT-ci therapist responds to communication from patients or caregivers for quick intervention. The goal is to coordinate a new or modified treatment plan to contend with the changes and seek to reestablish a new steady state of optimized functioning commensurate with the identified patient's current capabilities. This long-term follow-up will conceivably continue until the identified patient dies or is placed in a long-term care facility (if he or she remains in the same locale).

Once steady state is achieved, the IPT-ci therapist also has the unique opportunity to diplomatically raise important issues for consideration such as advanced directives, last will and testaments, end of life preparations, and so on. Furthermore, should cognitive function decline in the identified patient enough to warrant placement in a long-term care facility, the therapist can prepare the patient and caregiver ahead of time with thoughtful contingencies that help avoid facing a crisis that requires immediate and hasty action. The role of the IPT-ci therapist is one of an educator and facilitator for the identified patient and the caregiver(s). In addition, by virtue of the time the therapist has spent getting to know and following the patient and caregivers long term, the therapist may also serve as the most knowledgeable healthcare team member to liaison with the rest of the health care team for input into decision making.

Using IPT-ci as a Training Paradigm for a Geriatric Mental Health Specialist

The IPT-ci therapist may come from a social work, nursing, psychology, or medicine backgrounds. With the predicted boom in population of older Americans, many of whom will require such services as they age, using IPT-ci as a basis for a new training program designed for managing these needs in the primary care setting is an idea whose time may have come. Perhaps the ideal strategy for the training of IPT-ci therapists could be nested within social work training programs such that IPT-ci therapists would also be knowledgeable about community resources and be able to connect caregivers and the identified patient to any of a variety of services within their locale. It could be argued that the IPT-ci model of care is an excellent template for conceptualizing commonly encountered problems, building rapport with all parties, and effectively intervening to optimize the psychosocial care component for those older patients with cognitive impairment or dementia as well as depression. The added skills and knowledge specific to a social work degree (e.g., how to access a panoply of services) would allow a trained individual at the master's degree level to act independently within a primary care setting. Such a Geriatric Mental Health Specialist (GMHS) would operate in a similar way that depression care specialists were utilized successfully in the PROSPECT (Bruce et al., 2004) and IMPACT studies (Unutzer et al., 2002) for optimizing the treatment of depression. In those studies, the depression care specialist was backed up by off-site psychiatric consultation for difficult cases by telephone or directly with the patient face to face as indicated for good over all care. In a primary care office, of course, some family medicine or internal medicine physicians may be comfortable prescribing and adjusting psychotropic medications for depression and dementia but most are eager to have the consultative services of a well-trained psychiatrist (preferably a geriatric psychiatrist) for these patients when they are uncertain how best to manage them.

Finally, although efficacy studies need to be completed to confirm the face validity of IPT-ci, the principles of IPT-ci could also be applied in a group setting in the following variations. As the main features of IPT-ci are its techniques for engaging and educating both identified patients with cognitive impairment or depression and caregivers or other concerned persons and then attempting to bridge any gaps in understanding or address role disputes that might exist between the two, an IPT-ci group might do so for three to four patients and interested family/caregivers simultaneously. Choosing who would attend would, of course, require some judgment such that the severity level of cognitive impairment among the identified patients was fairly matched and such groups would probably target those in the early stages of cognitive decline as that is where the steepest learning curve resides. It would be reasonable to have a separate group just for caregivers in a similar way that individual sessions are encouraged in IPT-ci in order to allow the caregivers to speak freely of their concerns and frustrations without worrying that their comments might be misinterpreted by the identified patient or to allay fears about making them feel guilty for causing distress in others, and so on. It has been traditional in Alzheimer Association Support Groups and other similar organizations to hold two simultaneous group sessions, one for caregivers and the other for identified patients, with agenda tailored to the individual needs of each group. In the patient group, the topics might include how to cope with declining or lost abilities and how to seek replacements for the satisfaction they once brought or how to reasonably advocate for one's own autonomy when a well-meaning caregiver feels overbearing or overprotective.

These group IPT-ci endeavors could expand the scope and reach of an IPT-ci therapist on a larger scale than the therapist otherwise would have time to attend to. Furthermore, just like in other therapeutic groups, the astute IPT-ci therapist would be making mental notes about any individual patient or caregiver issue that came to light in the group setting that could be further addressed on a one-to-one basis at a later time. In this way, group and individual IPT-ci interventions could complement each other.

The group format of IPT-ci may also be an ideal one to use within long-term care settings.

Further Research

Although an expanding body of literature exists to document the efficacy of IPT for use in cognitively intact individuals, more research is needed to confirm the face validity of IPT-ci. Those who have been working with this population undoubtedly already carry out similar interventions even though background training and perspective may differ. Clearly there is no single correct way to do this work.

Designing research trials with IPT-ci are fraught with several difficulties. Identified patients can present with concomitant medical complexity. Caregivers would also need to consent to be evaluated for their role in any intervention outcomes. For those with a diagnosis of major depression, the confound of antidepressant medication administration adds complexity, for example, identified patients often arrive partially treated with such drugs or may have a history of intolerance to certain ones. Any attempt to draw conclusions about the relative beneficial effects of drug versus IPT-ci would ideally require standardized approaches to pharmacotherapy and such approaches are difficult to implement in primary care settings. Furthermore, any comparison of IPT-ci to "usual care" still must contend with the question of the comparability of whatever psychotropic medications are used in both groups of patients as well as the difficulties of making precise assessments of cognitive status.

It is the author's sincere hope that the dissemination of the principles of IPT-ci set forth in this publication will be of immediate usefulness to those working with this population but also that various groups will find additional ways to meaningfully test its efficacy and usefulness in a variety of settings.

Appendix

Montreal Cognitive Assessment (MoCA) Administration and Scoring Instructions

The Montreal Cognitive Assessment (MoCA) was designed as a rapid screening instrument for mild cognitive dysfunction. It assesses different cognitive domains: attention and concentration, executive functions, memory, language, visuoconstructional skills, conceptual thinking, calculations, and orientation. Time to administer the MoCA is approximately 10 minutes. The total possible score is 30 points; a score of 26 or above is considered normal.

1. **Alternating Trail Making**:
   Administration: The examiner instructs the subject: "Please draw a line, going from a number to a letter in ascending order. Begin here [point to (1)] and draw a line from 1 then to A then to 2 and so on. End here [point to (E)]."
   Scoring: Allocate one point if the subject successfully draws the following pattern: 1 –A- 2- B- 3- C- 4- D- 5- E, without drawing any lines that cross. Any error that is not immediately self-corrected earns a score of 0.

2. **Visuoconstructional Skills (Cube)**:
   Administration: The examiner gives the following instructions, pointing to the cube: "Copy this drawing as accurately as you can, in the space below".
   Scoring: One point is allocated for a correctly executed drawing.
   - Drawing must be three-dimensional
   - All lines are drawn
   - No line is added
   - Lines are relatively parallel and their length is similar (rectangular prisms are accepted)

   A point is not assigned if any of the above-criteria are not met.

3. **Visuoconstructional Skills (Clock)**:
   Administration: Indicate the right third of the space and give the following instructions: "Draw a clock. Put in all the numbers and set the time to 10 after 11."
   Scoring: One point is allocated for each of the following three criteria:
   - Contour (1 pt.): the clock face must be a circle with only minor distortion acceptable (e.g., slight imperfection on closing the circle);
   - Numbers (1 pt.): all clock numbers must be present with no additional numbers; numbers must be in the correct order and placed in the approximate quadrants on the clock face; Roman numerals are acceptable; numbers can be placed outside the circle contour;

- Hands (1 pt.): there must be two hands jointly indicating the correct time; the hour hand must be clearly shorter than the minute hand; hands must be centred within the clock face with their junction close to the clock centre. A point is not assigned for a given element if any of the above-criteria are not met.

4. Naming:
Administration: Beginning on the left, point to each figure and say: "Tell me the name of this animal."
Scoring: One point each is given for the following responses: (1) camel or dromedary, (2) lion, (3) rhinoceros or rhino.

5. Memory:
Administration: The examiner reads a list of 5 words at a rate of one per second, giving the following instructions: "This is a memory test. I am going to read a list of words that you will have to remember now and later on. Listen carefully. When I am through, tell me as many words as you can remember. It doesn't matter in what order you say them." Mark a check in the allocated space for each word the subject produces on this first trial. When the subject indicates that (s)he has finished (has recalled all words), or can recall no more words, read the list a second time with the following instructions: "I am going to read the same list for a second time. Try to remember and tell me as many words as you can, including words you said the first time." Put a check in the allocated space for each word the subject recalls after the second trial.
At the end of the second trial, inform the subject that (s)he will be asked to recall these words again by saying, "I will ask you to recall those words again at the end of the test."
Scoring: No points are given for Trials One and Two.

6. Attention:
Forward Digit Span: Administration: Give the following instruction: "I am going to say some numbers and when I am through, repeat them to me exactly as I said them." Read the five number sequence at a rate of one digit per second.
Backward Digit Span: Administration: Give the following instruction: "Now I am going to say some more numbers, but when I am through you must repeat them to me in the backwards order." Read the three number sequence at a rate of one digit per second.
Scoring: Allocate one point for each sequence correctly repeated (N.B.: the correct response for the backwards trial is 2-4-7).
Vigilance: Administration: The examiner reads the list of letters at a rate of one per second, after giving the following instruction: "I am going to read a sequence of letters. Every time I say the letter A, tap your hand once. If I say a different letter, do not tap your hand."
Scoring: Give one point if there is zero to one errors (an error is a tap on a wrong letter or a failure to tap on letter A).
Serial 7s: Administration: The examiner gives the following instruction: "Now, I will ask you to count by subtracting seven from 100, and then, keep subtracting seven from your answer until I tell you to stop." Give this instruction twice if necessary.
Scoring: This item is scored out of 3 points. Give no (0) points for no correct subtractions, 1 point for one correction subtraction, 2 points for two-to-three

correct subtractions, and 3 points if the participant successfully makes four or five correct subtractions. Count each correct subtraction of 7 beginning at 100. Each subtraction is evaluated independently; that is, if the participant responds with an incorrect number but continues to correctly subtract 7 from it, give a point for each correct subtraction. For example, a participant may respond
"92 − 85 − 78 − 71 − 64" where the "92" is incorrect, but all subsequent numbers are subtracted correctly. This is one error and the item would be given a score of 3.

7. **Sentence repetition**:
   Administration: The examiner gives the following instructions: "I am going to read you a sentence. Repeat it after me, exactly as I say it [pause]: I only know that John is the one to help today." Following the response, say: "Now I am going to read you another sentence. Repeat it after me, exactly as I say it [pause]: The cat always hid under the couch when dogs were in the room."
   Scoring: Allocate 1 point for each sentence correctly repeated. Repetition must be exact. Be alert for errors that are omissions (e.g., omitting "only", "always") and substitutions/additions (e.g., "John is the one who helped today"; substituting "hides" for "hid", altering plurals, etc.).

8. **Verbal fluency**:
   Administration: The examiner gives the following instruction: "Tell me as many words as you can think of that begin with a certain letter of the alphabet that I will tell you in a moment. You can say any kind of word you want, except for proper nouns (like Bob or Boston), numbers, or words that begin with the same sound but have a different suffix, for example, love, lover, loving. I will tell you to stop after one minute. Are you ready? [Pause] Now, tell me as many words as you can think of that begin with the letter F. [time for 60 sec]. Stop."
   Scoring: Allocate one point if the subject generates 11 words or more in 60 sec. Record the subject's response in the bottom or side margins.

9. **Abstraction**:
   Administration: The examiner asks the subject to explain what each pair of words has in common, starting with the example: "Tell me how an orange and a banana are alike." If the subject answers in a concrete manner, then say only one additional time: "Tell me another way in which those items are alike." If the subject does not give the appropriate response (fruit), say, "Yes, and they are also both fruit." Do not give any additional instructions or clarification.

   After the practice trial, say: "Now, tell me how a train and a bicycle are alike." Following the response, administer the second trial, saying: "Now tell me how a ruler and a watch are alike." Do not give any additional instructions or prompts.
   Scoring: Only the last two item pairs are scored. Give 1 point to each item pair correctly answered. The following responses are acceptable:
   Train-bicycle = means of transportation, means of travelling, you take trips in both;
   Ruler-watch = measuring instruments, used to measure.
   The following responses are not acceptable: Train-bicycle = they have wheels; Ruler-watch = they have numbers.

10. **Delayed recall**:
    Administration: The examiner gives the following instruction: "I read some words to you earlier, which I asked you to remember. Tell me as many of those words as you can remember. Make a check mark (✓) for each of the words

correctly recalled spontaneously without any cues, in the allocated space.
Scoring: Allocate 1 point for each word recalled freely without any cues.

> **Optional**:
> Following the delayed free recall trial, prompt the subject with the semantic category cue provided below for any word not recalled. Make a check mark (✓) in the allocated space if the subject remembered the word with the help of a category or multiple-choice cue. Prompt all non-recalled words in this manner. If the subject does not recall the word after the category cue, give him/her a multiple choice trial, using the following example instruction, "Which of the following words do you think it was, NOSE, FACE, or HAND?"
> Use the following category and/or multiple-choice cues for each word, when appropriate:
> FACE: category cue: part of the body   multiple choice: nose, face, hand
> VELVET: category cue: type of fabric   multiple choice: denim, cotton, velvet
> CHURCH: category cue: type of building   multiple choice: church, school, hospital
> DAISY: category cue: type of flower   multiple choice: rose, daisy, tulip
> RED: category cue: a colour   multiple choice: red, blue, green
> Scoring: No points are allocated for words recalled with a cue. A cue is used for clinical information purposes only and can give the test interpreter additional information about the type of memory disorder. For memory deficits due to retrieval failures, performance can be improved with a cue. For memory deficits due to encoding failures, performance does not improve with a cue.

11. **Orientation**:
    Administration: The examiner gives the following instructions: "Tell me the date today". If the subject does not give a complete answer, then prompt accordingly by saying: "Tell me the [year, month, exact date, and day of the week]." Then say: "Now, tell me the name of this place, and which city it is in."
    Scoring: Give one point for each item correctly answered. The subject must tell the exact date and the exact place (name of hospital, clinic, office). No points are allocated if subject makes an error of one day for the day and date.
    TOTAL SCORE: Sum all subscores listed on the right-hand side. Add one point for an individual who has 12 years or fewer of formal education, for a possible maximum of 30 points. A final total score of 26 and above is considered normal.

Copyright© 2003, 2004, 2005, 2006, 2007, 2008 The Montreal Cognitive Assessment. All rights reserved.

**For more information or feedback on MoCA©
contact Dr Z. Nasreddine at info@mocatest.org**

References

The American Journal of Geriatric Psychiatry. (2008a). NeuroPsychiatry Review [On-line]. Available: http://neuropsychiatryreviews.com/dec/00/npr_dec00_rtms.html

The American Journal of Geriatric Psychiatry. (2008b). Mayo Clinic.com [On-line]. Available: http://www.mayoclinic.com/health/vagus-nerve-stimulation/MH00113

The American Journal of Geriatric Psychiatry. (2008c). Mayo Clinic.com [On-line]. Available: http://www.mayoclinic.com/health/vagus-nerve-stimulation/MH00113

Alexopoulos, G. S. (2003). Vascular disease, depression, and dementia. *Journal of the American Geriatrics Society, 5*, 1178–1190.

Alexopoulos, G. S., Kiosses, D. N., Heo, M., Murphy, C. F., Shanmugham, B., & Gunning-Dixon, F. (2005). Executive dysfunction and the course of geriatric depression. *Biological Psychiatry, 58*, 204–210.

Alexopoulos, G. S., Meyers, B. S., Young, R. C., Kakuma, T., Silbersweig, D., & Charlson, M. (1997). Clinically defined vascular depression. *American Journal of Psychiatry, 154*, 562–565.

Alexopoulos, G. S., Raue, P., & Arean, P. A. (2003). Problem-solving therapy versus supportive therapy in geriatric major depression with executive dysfunction. *American Journal of Geriatric Psychiatry, 11*, 46–52.

American Psychiatric Association. (2000). *Diagnostic and statistical manual of mental disorders* (4th ed., Text Revision). Washington, DC: Author.

Arean, P. A., & Alexopoulos, G. S. (2007). Psychosocial interventions for mental illness in late-life. *International Journal of Geriatric Psychiatry, 22*, 99–100.

Arean, P. A., Perri, M. G., Nezu, A. M., Schein, R. L., Christopher, F., & Joseph, T. X. (1993). Comparative effectiveness of social problem-solving therapy and reminiscence therapy as treatments for depression in older adults. *Journal of Consulting and Clinical Psycholoygy, 61*, 1003–1010.

Badner, J. A., & Gershon, E. S. (2002). Meta-analysis of whole-genome linkage scans of bipolar disorder and schizophrenia. *Molecular Psychiatry, 7*, 405–411.

Baldwin, R., Jeffries, S., Jackson, A., Sutcliffe, C., Thacker, N., Scott, M., et al. (2004). Treatment response in late-onset depression: relationship to neuropsychological, neuroradiological and vascular risk factors. *Psychological Medicine, 34*, 125–136.

Barnes, D. E., Alexopoulos, G. S., Lopez, O. L., Williamson, J. D., & Yaffe, K. (2004). Depressive symptoms, vascular disease, and mild cognitive impairment: findings from the Cardiovascular Health Study. *Archives of General Psychiatry, 63*, 273–279.

Bertram, L., & Tanzi, R. E. (2005). The genetic epidemiology of neurodegenerative disease. *Journal of Clinical Investigation, 115*, 1449–1457.

Bharucha, A. J., Dew, M. A., Miller, M. D., Borson, S., & Reynolds, C. F. (2006). Psychotherapy in long-term care settings: A review. *Journal of the American Medical Directors Association, 7,* 568–580.

Bouwman, F. H., van der Flier, W. M., Schoonenboom, N. S., van Elk, E. J., Kok, A., Rijmen, F., et al. (2007). Longitudinal changes of CSF biomarkers in memory clinic patients. *Neurology, 69,* 1006–1011.

Bruce, J. M., Bhalla, R., Westervelt, H. J., Davis, J., Williams, V., & Tremond, G. (2008). Neuropsychological correlates of self-reported depression and self-reported cognition among patients with mild cognitive impairment. *Journal of Geriatric Psychiatry and Neurology, 21,* 34–40.

Bruce, M. L., Ten Have, T. R., Reynolds, C. F., Katz, I. I., Schulberg, H. C., Mulsant, B. H., et al. (2004). Reducing suicidal ideation and depressive symptoms in depressed older primary care patients: A randomized controlled trial. *Journal of the American Medical Association, 291,* 1081–1091.

Burdick, D. J., Rosenblatt, A., Samus, Q. M., Steele, C., Baker, A., Harper, M., et al. (2005). Predictors of functional impairment in residents of assisted-living facilities: the Maryland Assisted Living study. *The Journals of Gerontology. Series B, Psychological Sciences and Social Sciences, 60,* 258–264.

Burgio, L., Corcoran, M., Lichstein, L. L., Nichols, L., Czaja, S., Gallagher-Thompson, D., et al. (2001). Judging outcomes in psychosocial interventions for dementia caregivers: The problem of treatment implementation. *The Gerontologist, 41,* 481–490.

Burns, A., & O'Brien, J. (2006). Clinical practice with anti-dementia drugs: a consensus statement from British Association for Psychopharmacology. *Journal of Psychopharmacology, 20*(6), 732–755.

Butters, M. A., Whyte, E. M., Nebes, R. D., Begley, A. E., Dew, M. A., Mulsant, B. H., et al. (2004). The nature and determinants of neuropsychological functioning in late-life depression. *Archives of General Psychiatry, 61,* 587–595.

Buysse, D. J., Hubbard, R., Ombao, H., Houck, P., & Monk, T. H. (2001). Sleep-dependent and circadian influences on normal nocturnal sleep. *Sleep, 24*(Abstract Supplement), A1.

Cairney, J., Corna, L. M., Veldbuizen, S., Herrmann, N., & Streimer, D.L. (2008). Comorbid depression and anxiety in later life: Patterns of association, subjective well-being, and impairment. *American Journal of Geriatric Psychiatry, 16,* 201–208.

Carreira, K., Miller, M. D., Frank, E., Houck, P. R., Morse, J. Q., Dew, M. A. et al. (2008). A controlled evaluation of monthly maintenance Interpersonal Psychotherapy in late-life depression with varying levels of cognitive performance. *International Journal of Geriatric Psychiatry, 23,* 1110–1113.

Cummings, J. L., Schneider, E., Tariot, P. N., & Graham, S. M. (2006). Behavioral effects of memantine in Alzheimer disease patients receiving donepezil treatment. *Neurology, 67,* 57–63.

Dew, M. A., Hoch, C. C., Buysse, D. J., Monk, T. H., Begley, A. E., Houck, P. R., et al. (2003). Healthy older adults' sleep predicts all-cause mortality at 4 to 19 years of follow-up. *Psychosomatic Medicine, 65,* 63–73.

Driscoll, H. C., Karp, J. F., Dew, M. A., & Reynolds, C. F. (2007). Getting better, getting well: Understanding and managing partial and non-response to pharmacological treatment of non-psychotic major depression in old age. *Drugs & Aging, 24,* 801–814.

Driscoll, H. C., Serody, L., Patrick, S., Maurer, J., Bensasi, S., Houck, P. R., et al. (2008). Sleeping well, aging well: A descriptive and cross-sectional study of sleep in "successful agers" 75 and older. *American Journal of Geriatric Psychiatry, 16,* 74–82.

Erikson, E. H. (1950). *Childhood and society.* New York: Norton.

Feil, N. (2002). *The validation breakthrough.* Baltimore: Health Professions Press, Inc.

Feldman, H., Gauthier, S., Hecker, J., Vellas, B., Subbiah, P., & Whalen, E. (2001). A 24-week, randomized, double-blind study of donepezil in moderate to severe Alzheimer's disease. *Neurology, 57,* 613–620.

Folstein, M. F., Folstein, S. W., & McHugh, P. R. (1975). Mini-mental state: A practical method for grading the cognitive state of patients for the clinician. *Journal of Psychiatric Research, 12,* 189–198.

Friedman, L. J. (1999). *Identity's architect: A biography of Erik H. Erikson.* New York: Scribners.

Gabryelewicz, T., Styczynska, M., Pfeffer, A., Wasiak, B., Barczak, A., Luczywek, E., et al. (2004). Prevalence of major and minor depression in elderly persons with mild cognitive impairment—MADRS factor analysis. *International Journal of Geriatric Psychiatry, 12,* 1168–1172.

Gallagher-Thompson, D. & Steffen, A. M. (1982). Treatment of major depressive disorder in older adult outpatients with brief psychotherapies. *Psychotherapy: Theory, Research, Practice, Training, 19,* 482–490.

Gallagher-Thompson, D., Steffen, A.M., & Thompson, L.W. (Eds.). (2008). *Handbook of behavioral and cognitive therapies with older adults.* New York: Springer.

Gallagher-Thompson, D., & Thompson, L. W. (2006). The journal of aging and mental health. *Clinical Gerontologist, 30,* 1–110.

Ganguli, M., Du, Y., Rodriguez, E. G., Mulsant, B. H., McMichael, K. A., Vander Bilt, J., et al. (2006). Discrepancies in information provided to primary care physicians by patients with and without dementia: The Steel Valley Seniors Survey. *American Journal of Geriatric Psychiatry, 14,* 446–455.

George, L. K., & Gwyther, L. P. (1986). Caregiver well-being: A multidimensional examination of family caregivers of demented adults. *The Gerontologist, 26,* 253–259.

Gitlin, L., Belle, S. H., Burgio, L., Czaja, S. J., Mahoney, D., Gallagher-Thompson, D., et al. (2003). Effects of multi-component interventions on caregiver burden and depression: The REACH multi-site initiative at 6 months follow-up. *Psychology and Aging, 18,* 361–374.

Goldapple, K., Segal, Z., Garson, C., Lau, M., Bieling, P., Kennedy, S. et al. (2004). Modulation of cortical-limbic pathways in major depression: treatment-specific effects of cognitive behavior therapy. *Archives of General Psychiatry, 61,* 34–41.

Grossman, M. (2002). Frontotemporal dementia: A review. *Journal of International Neuropsychological Society, 8(4),* 566–583.

Gwyther, L. P. (2004). Working with the family of the older adult. In D.G. Blazer, D. C. Steffens, & E. W. Busse (Eds.), *Textbook of Geriatric Psychiatry* (3rd ed., pp. 459–471). Washington DC: American Psychiatric Publishing Inc.

Halliday, G., Robinson, S. R., & Shepherd, C. (2000). Alzheimer's disease and inflammation: A review of cellular and therapeutic mechanisms. *Clinical and Experimental Pharmacology & Physiology, 27,* 1–8.

References

Heisel, M. J., Links, P. S., Conn, D., vam Reekum, R., & Flett, G. L. (2007). Narcissistic personality and vulnerability to late-life suicidality. *The American Journal of Geriatric Psychiatry, 15*, 734–741.

Hinrichsen, G. A. (2007). Series X—Clinical geropsychology. Ref Type: Video Recording.

Hinrichsen, G. A., & Clougherty, K. F. (2006). *Interpersonal psychotherapy for depressed older adults* (1st ed.). Washington, DC: APA.

Hinrichsen, G. A., & Pollack, S. (1997). Expressed emotion and the course of late-life depression. *Journal of Abnormal Psychology, 106*, 336–340.

Hirschhorn, J. N., Lohmueller, K., Byrne, E., & Hirschhorn, K. (2002). A comprehensive review of genetic association studies. *Genetics in Medicine, 4*, 45–61.

Hoare, C. H. (2002). *Erikson on development in adulthood: New insights from the unpublished papers*. New York: Oxford University Press.

Jefferson, A. L., Byerly, L. K., Vanderhill, S., Lambe, S., Wong, S., Ozonoff, A., et al. (2008). Characterization of activities of daily living in individuals with mild cognitive impairment. *American Journal of Geriatric Psychiatry, 16*, 375–383.

Jeste, D. V., Lohr, J. B., & Goodwin, F. K. (1988). Neuroanatomical studies of major affective disorders. A review and suggestions for further research. [Review] [131 refs]. *The British Journal of Psychiatry: The Journal of Mental Science, 153*, 444–459.

Kalayam, B., & Alexopoulos, G. S. (1999). Prefrontal dysfunction and treatment response in geriatric depression. *Archives of General Psychiatry, 56*, 713–718.

Karp, J. F., Weiner, D., Seligman, K., Butters, M., Miller, M., Frank, E. et al. (2005). Body pain and treatment response in late-life depression. *American Journal of Geriatric Psychiatry, 13*, 188–194.

Kaufer, D. I. (2002). Cholinesterase-inhibitor Therapy for dementia: Novel clinical substrates and mechanisms for treatment response. *CNS Spectrums, 7*, 742–750.

Kessler, R. C., Berglund, P., Demler, O., Jin, R., Merikangas, K. R., & Walters, E. E. (2005). Lifetime prevalence and age-of-onset distributions of DSM-IV disorders in the National Comorbidity Survey Replication. *Archives of General Psychiatry, 62*, 593–602.

Klerman, G. L., Weissman, M. M., Rounsaville, B. J., & Chevron, E. (1984). *Interpersonal psychotherapy of depression*. New York: Academic Press, Basic Books Inc.

Kupfer, D., Horner, M.S., Brent, D., Lewis, D., Reynolds, C., Thase, M., Travis, M., & Horner, M. X. (Eds.). (2008). *Oxford American handbook of psychiatry*. New York: Oxford University Press.

Lanctot, K. L., Herrmann, N., Yau, K. K., Khan, L. R., Liu, B. A., Loulou, M. M., et al. (2003). Efficacy and safety of cholinesterase inhibitors in Alzheimer's disease: A meta-analysis. *Canadian Medical Association Journal, 169*, 557–564.

Lawrence, N. S., Williams, A. M., Surguladze, S., Giampietro, V., Brammer, M. J., Andrew, C., et al. (2004). Subcortical and ventral prefrontal cortical neural responses to facial expressions distinguish patients with bipolar disorder and major depression. *Biological Psychiatry, 55*, 578–587.

Lenze, E. (2007). Anxiety depression in the elderly. *Perspectives in Psychiatry, 2*, 3–6.

Linehan, M. M. (1993). *Cognitive-behavioral treatment of borderline personality disorder*. New York: Guilford Press.

Lockwood, K. A., Alexopoulos, G. S., Kakuma, T., & Van Gorp, W. G. (2000). Subtypes of cognitive impairment in depressed older adults. *American Journal of Geriatric Psychiatry, 8*, 201–208.

Lu, Y. F., Austrom, M. G., Perkins, S. M., Bakas, T., Farlow, M. R., He, F., et al. (2007). Depressed mood in informal caregivers of individuals with mild cognitive impairment. *American Journal of Alzheimer's Disease and Other Dementias, 22*, 273–285.

Lyketsos, C. G., Lopez, O., Jones, B., Fitzpatrick, A. L., Breitner, J., & DeKosky, S. (2002). Prevalence of neuropsychiatric symptoms in dementia and mild cognitive impairment: results from the cardiovascular health study. *Journal of the American Medical Association, 288*, 1475–1483.

Lynch, T. R. (2000). Treatment of elderly depression with personality disorder comorbidity using dialectical behavior therapy. *Cognitive and Behavioral Practice, 7(4)*, 468–477.

Lynch, T. R., Cheavens, J. S., Cukrowicz, K. C., Thorp, S. R., Bronner, L., & Beyer, J. (2007). Treatment of older adults with co-morbid personality disorder and depression: A dialectical behavior therapy approach. *International Journal of Geriatric Psychiatry, 22(7)*, 702–703.

Lynch, T. R., Morse, J. Q., Mendelson, T., & Robins, C. J. (2003). Dialectical behavior therapy for depressed older adults: A randomized pilot study. *The American Journal of Geriatric Psychiatry, 11*, 33–45.

Lyness, J. M. (2002). The cerebrovascular model of depression in late life. *CNS Spectrums, 7*, 712–715.

Mackin, R. S., & Arean, P. A. (2007). Cognitive and psychiatric predictors of medical treatment adherence among older adults in primary care clinics. *International Journal of Geriatric Psychiatry, 22*, 55–60.

Maletic, V., Robinson, M., Oakes, T., Lyengar, S., Ball, S. G., & Russell, J. (2007). Neurobiology of depression: an integrated view of key findings. *International Journal of Clinical Practice, 61*, 2030–2040.

Mayberg, H. S. (2003). Modulating dysfunctional limbic-cortical circuits in depression: Towards development of brain-based algorithms for diagnosis and optimised treatment. *British Medical Bulletin, 65*, 193–207.

Mayberg, H. S. (2006). Defining neurocircuits in depression: Strategies toward treatment selection based on neuroimaging phenotypes. *Psychiatric Annals, 36*, 259–268.

Mayberg, H., Lozano, A. M., Voon, V., McNeely, H. E., Seminowicz, D., Hamani, C. et al. (5 A.D.). Deep Brain Stimulation for Treatment-Resistant Depression. *Neuron, 45*, 651–660.

Miller, M. D., Lenze, E. J., Dew, M. A., Whyte, E., Weber, E., Begley, A. E., et al. (2002). Effect of cerebrovascular risk factors on depression treatment outcome in later life. *American Journal of Geriatric Psychiatry, 10*, 592–598.

Miller, M. D., & Reynolds, C. F. (2002). *Living Longer Depression Free: A Family Guide to Recognizing, Treating and Preventing Depression in Later Life*. Baltimore, MD: The Johns Hopkins University Press.

Mullan, M., Crawford, F., Axelman, K., & Houlden, H. (1992). A pathogenic mutation for probable Alzheimer's disease in the APP gene at the N-terminus of amyloid. *Nature Genetics, 1*, 345–347.

Nasreddine, Z. S., Phillips, N. A., Bédirian, V., Charbonneau, S., Whitehead, V., Collin, I. et al. (2005). The Montreal Cognitive Assessment, MoCA: A brief screening tool for mild cognitive impairment. *Journal of the American Geriatrics Society, 53*, 695–699.

Nezu, A. M. (1987). A problem-solving formulation of depression: a literature review and proposal of pluralistic model. *Clinical Psychology Review, 7*, 121–144.

Oslin, D. W., Katz, I. R., Edell, W. S., & Ten Have, T. R. (2000). Effects of alcohol consumption on the treatment of depression among elderly patients. *American Journal of Psychiatry, 8*, 215–220.

Palmer, K., Backman, L., Winblad, B., & Fratiglioni, L. (2008). Mild cognitive impairment in the general population: occurrence and progression to Alzheimer disease. *American Journal of Geriatric Psychiatry, 16*, 603–611.

Palmer, K., Berger, A. K., Monastero, R., Winblad, B., Backman, L., & Fratiglioni, L. (2007). Predictors of progression from mild cognitive impairment to Alzheimer disease. *Neurology, 68*, 1596–1602.

Petersen, R. C., & Negash, S. (2008). Mild cognitive impairment: An overview. *CNS Spectrums, 13*, 45–53.

Petersen, R. C., Smith, G. E., Waring, S. C., Ivnik, R. J., Tangalos, E. G., & Kokmen, E. (1999). Mild cognitive impairment: Clinical characterization and outcome. *Archives of Neurology, 56*, 303–308.

Pilkonis, P. A., & Frank, E. (1988). Personality pathology in recurrent depression: Nature, prevalence, and relationship to treatment response. *American Journal of Psychiatry, 145*, 435–441.

Post, E. P., Miller, M., & Schulberg, H. C. (2008). Treating depression in older primary care patients with interpersonal psychotherapy. *Geriatrics, 63*, 18–28.

Qaseem, A., Snow, V., Cross, J. T., Forciea, M. A., Hopkins, R., Shekelle, P., et al. (2008). Current pharmacologic treatment of dementia: A clinical practice guideline from the American College of Physicians and the American Academy of Family Physicians. *Annals of Internal Medicine, 148*, 370–378.

Reding, M., Haycox, J., & Blass, J. (1985). Depression in patients referred to a dementia clinic. A three-year prospective study. *Archives of Neurology, 42*, 894–896.

Reynolds, C. F., Dew, M. A., Pollock, B. G., Mulsant, B. H., Frank, E., Miller, M. D., et al. (2006). Maintenance treatment of major depression in old age. *New England Journal of Medicine, 354*, 1130–1138.

Reynolds, C. F., Frank, E., Perel, J. M., Imber, S. D., Cornes, C., Miller, M. D., et al. (1999). Nortriptyline and interpersonal psychotherapy as maintenance therapies for recurrent major depression: A randomized controlled trial in patients older than 59 years. *Journal of the American Medical Association, 281*, 39–45.

Ringman, J. M. (2006). Donepezil improves cognitive function and reduces decline in activities of daily living in people with severe Alzheimer's disease. *Evidence-Based Mental Health, 9*, 104.

Royall, D. R., Mahurin, R. K., & Gray, K. F. (1992). Bedside assessment of executive cognitive impairment: The Executive Interview. *Journal of the American Geriatrics Society, 40*, 1221–1226.

Rozzini, L., Chilovi, B. V., Trabucchi, M., & Padovani, A. (2008). Predictors of progression from mild cognitive impairment to Alzheimer disease. *Neurology, 70*, 735–736.

Sakauye, K. (2008). *Geriatric psychiatry basics* (1st ed.). New York: W.W. Norton.

Schulberg, H. C., Post, E. P., Raue, P. J., Ten Have, T., Miller, M., & Bruce, M. L. (2007). Treating late-life depression with interpersonal psychotherapy in the primary care sector. *International Journal of Geriatric Psychiatry, 22*, 106–114.

Schulz, R., & Patterson, T. L. (2004). Caregiving in geriatric psychiatry. *American Journal of Geriatric Psychiatry, 12*, 234–237.

Shear, M. K., Frank, E., Houck, P. R., & Reynolds, C. F. (2005). Treatment of complicated grief: A randomized controlled trial. *Journal of the American Medical Association, 293*, 2601–2608.

Sheline, Y. I., Mintun, M. A., Barch, D. M., Wilkins, C., Snyder, A. Z., & Moerlein, S. M. (2004). Decreased hippocampal 5-HT(2A) receptor binding in older depressed patients using [18F]altanserin positron emission tomography. *Neuropsychopharmacology, 29*, 2235–2241.

Skultety, K. M., & Rodriguez, R. L. (2008). Treating geriatric depression in primary care. *Current Psychiatry Reports, 10*, 44–49.

Smoller, J. W., & Gardner-Schuster, E. (2007). Genetics of bipolar disorder. *Current Psychiatry Reports, 9*, 504–511.

Spar, J. E., & LaRue, A. (2006a). *Clinical manual of geriatric psychiatry* (1st ed.) Washington, DC: American Psychiatric Publishing, Inc.

Spar, J. E., & LaRue, A. (2006b). Other Dementias and Delirium. In Spar, J. E. & LaRue, A. (Eds.), *Clinical manual of geriatric psychiatry* (1st ed., pp. 229–272). Arlington: American Psychiatric Publishing Inc.

Steffens, D. C., Hays, J. C., & Krishnan, K. R. (1999). Disability in geriatric depression. *American Journal of Geriatric Psychiatry, 7*, 34–40.

Stein, D. J. (2008). Emotional regulation: Implications for the psychobiology of psychotherapy. *CNS Spectrums, 13*, 195–198.

Stuart, S. & Robertson, M. (2003). *Interpersonal Psychotherapy: A clinician's Guide.* London: Edward Arnold Ltd.

Sweet, R. A., Hamilton, R. L., Butters, M. A., Mulsant, B. H., Pollock, B. G., Lewis, D. A., et al. (2004). Neuropathologic correlates of late-onset major depression. *Neuropsychopharmacology, 29*, 2242–2250.

Tarsy, D., Vitek, J. L., Starr, P. A., & Okun, M. S. (Eds). (2008). *Deep brain stimulation in neurological and psychiatric disorders.* Totowa, NJ: Humana Press.

Teri, L., McKenzie, G., & LaFazia, D. (2005). Psychosocial treatment of depression in older adults with dementia. *Clinical Psychology: Science and Practice, 12*, 303–316.

Trepacz, P. T., Teague, G. B., & Lipowski, Z. J. (1985). Delirium and other organic mental disorders in a general hospital. *General Hospital Psychiatry, 7*, 101–106.

Tuokko, H., Morris, C., & Ebert, P. (2005). Mild cognitive impairment and everyday functioning in older adults. *Neurocase, 11*, 40–47.

Unutzer, J., Katon, W., Callahan, C. M., Williams, J. W., Hunkeler, E., Harpole, L., et al. (2002). Collaborative care management of late-life depression in the primary care setting: A randomized controlled trial. *Journal of the American Medical Association, 288*, 2836–2845.

Warden, D., Rush, A. J., Trivedi, M. H., Fava, M., & Wisniewski, S. R. (2007). The STAR*D Project results: A comprehensive review of findings. *Current Psychiatry Reports, 9*, 449–459.

Welchman, K. (2000). *Erik Erikson, his life, work, and significance.* Buckingham, UK: Open University Press.

Weiner, M. B. (1979). Caring for the elderly. Psychological aging: Aspects of normal personality and development in old age. Part II. Erik Erikson: Resolutions of psychosocial tasks. *The Journal of Nursing Care, 12*(5), 27–28.

Weissman, M. M. (1999). *Comprehensive guide to interpersonal psychotherapy.* New York: Basic Books.

Weissman, M. M., Markowitz, J. C., & Klerman, G. L. (2007). *Clinician's quick guide to interpersonal psychotherapy.* New York: Oxford University Press.

Welsh-Bohmer, K. A., Gearing, M., Saunders, A. M., Roses, A. D., & Mirra, S. (1997). Apolipoprotein E genotypes in a neuropathological series from the consortium to establish a registry for Alzheimer's disease. *Annals of Neurology, 42,* 319–325.

Whitwell, J. L., Shiung, M. M., Przybelski, S. A., Weigand, S. D., Knopman, D. S., Boeve, B. F. et al. (2008). MRI patterns of atrophy associated with progression to AD in amnestic mild cognitive impairment. *Neurology, 70,* 512–520.

Williamson, G., & Schulz, R. (1992). Physical illness and symptoms of depression among elderly outpatients. *Psychology and Aging, 7,* 343–351.

Winblad, B., & Poritis, N. (1999). Memantine in severe dementia: results of the 9M-Best study (Benefit and efficacy in severely demented patients during treatment with Memantine). *International Journal of Geriatric Psychiatry, 14,* 135–146.

Zubenko, G. S., Moossy, J., & Kopp, U. (1990). Neurochemical correlates of major depression in primary dementia. *Archives of Neurology, 47,* 209–214.

Index

Acetylcholine, 55, 56, 58, 63
Activities of daily living (ADL), 30, 35, 152
Adult child caregivers. *See also* caregivers; friends as caregivers; spouse caregivers
    abuse by, 110–113
    case vignettes, 104–106, 108–109
    conflicting other relationships of, 107–108
    dual stresses of, 104
    inadequacy of, 110–113
    intact spouse collaboration, 115–116
    IPT-ci therapist work with, 20, 26, 101, 109–110, 116, 135, 147, 150, 163
    micromanagement of parent, 31
    needs assessment of, 106–107
    role disputes with parents, 127–129
    role transition for, 18
    safety vs. parent autonomy choice, 35
    sharing of caregiving duties, 109–110
Advanced directives, 36, 88, 113t, 173–174, 176
Aging
    calming influence of, 133
    ED and, 79
    narcissistic PD and, 133
    physiologic changes of, 31–32, 42–43
    study of (*See* gerontology/geriatric medicine)
    successful, and role transitioning, 117t
Agitation
    as delirium symptom, 50, 51t
    as *DSM-IV-TR* depression criteria, 38t
    psychotropic medications for, 29, 90t
Alcohol/alcoholism
    additive/disinhibiting role of, 75, 79
    cognitive decline from, 138–142

    dementia syndrome and, 61–62
    depression from, 43
    ED and, 67, 150
    safety issues, 34
    withdrawal issues, 33, 52, 53
Alprazolam, 46, 52
Alzheimer Association Support Groups, 177
Alzheimer's disease, 3, 22, 24
    brain of, 54–56
        imaging research advances, 48
    depression and, 63
    drug treatments for, 55–56
    frontal group dementias vs., 54
    genetics of, 55
    MCI conversion to, 60
    prodromal period of symptoms, 59–60
    sense diminishment issues, 32
Alzheimer's Disease Association, 161t
Amnestic type MCI, 75, 118–119
Analgesics/anti-inflammatory agents and depression, 41t
Antibacterial/antifungal agents and depression, 41t
Anticholinesterases and depression, 41t
Antidepressant medication, 28, 37, 104, 105, 118. *See also* selective serotonin reuptake inhibitors
    IPT vs., 5, 8
    IPT with, 4–6, 8, 23, 69, 75, 100, 160
    talk therapy vs., 44, 45
    talk therapy with, 45
Antihypertensive agents and depression, 41t
Antineoplastic agents and depression, 41t
Antipsychotic agents and depression, 41t
Antiseizure medication, 33, 46
Antisocial personality disorder, 133

Index

Anxiety/anxiety disorders, 4
  case vignette, 39–40, 143–144, 166–168
  cognitive impairment and, 121
  co-occurring, 46
  generalized anxiety disorder, 46, 142
  OCD and, 143
  psychotropic medications for, 4, 29, 90
Appetite suppressants and depression, 42t
Aricept® (donepezil), 55, 56, 58, 96, 119, 161, 170–171
Aripiprazole, 46

Benefits of IPT, 11–12
Benzodiazepines, 46
Bipolar I (mania) disorder, 45
Bipolar II (hypomania) disorder, 45–46
Borderline personality disorder, 46–47, 134–135
Brain
  of Alzheimer victims, 54–56
  dementia autopsy studies, 59
  imaging studies
    CAT scans, 54, 62
    MRI scans, 54, 62, 81, 123
    SPECT scans of, 54, 62, 70
  vascular changes (of dementia), 57–58
B vitamins
  B1 (thiamine) deficiency, 61
  B12 deficiency, 59, 61

Caffeine, 33, 42t
Cardiac/antihypertensive agents and depression, 41t
Cardiopulmonary conditions and depression, 40t
Caregivers. *See also* adult child caregivers; spouse caregivers
  basic strategies for, 93t–94t
  coping skills/strategies of, 87
  denial of, 72, 113
  engagement of, 25–27
  factors affecting stress of, 148t
  friends as, 155–156
  lovers (paramour) as, 156–157
  nonfamily caregivers, 116
  role transitions of, 4, 25, 147–157
  talking to, about patient's dementia, 160–161, 161t
Case vignettes
  adult child caregiver, 104–106, 108–113
  complicated grief, 13–14
  daughter at mother's interview, 97
  delirium, 51–52
  dementia, 53–54, 59
  depression from medical condition, 39–40
  drug–drug interactions, 32–33
  emergency placement, 164–165
  friends as caregivers, 155–156
  future contingency planning, 166–171
  interpersonal deficit, 17–20, 130–132, 136–137
  long-term planning, 164–165
  loss of decision-making power, 75–78
  minimal cognitive impairment, 70
  role disputes (*See* role disputes, case vignettes)
  role transitions, 14–15, 111–113
  sporadic odd behavior/aggression, 68–70
  spouse becoming caregiver, 70–74
Cataract formation, 31
CAT scans, 54
Cholinergic enhancement agents, 28, 48, 55, 56, 58, 63, 78, 161t
Clinical management (CM), 4, 6–7
Cognitive-behavioral therapy (CBT), 11, 12, 44, 143, 154
Cognitive impairment. *See also* executive dysfunction (ED)
  abnormal grief in, 121–124
  additive demoralization from, 106
  depression and, 47–48
  early placement for (vignette), 164–165
  initial state treatments, 28
  IPT for, 24–25
  lack of insight into declining function, 130–132
  with predominant memory loss, 75, 118–119
  psychotropic medications for, 29
  as role transition, 25
  waning of decision-making power, 75–78
Cognitively intact geriatric patients (vignettes)
  depression in, 25
  IPT use with
    adjusting to retirement, 14–15
    complicated grief, 13–14
    dependent personality disorder, 17–20
    role dispute during adjustment to retirement, 15–17

Collaborative care model (of healthcare), 28, 48–49, 100
Contingency planning, 166–171
Contract (of IPT) with patients
  formation of, 9, 100–102, 160
  phase III termination, 10
  pressures created by, 11–12
Co-occurring anxiety, 46
Coping skills/strategies
  of caregivers, 87, 148t
  for depression avoidance, 25
  exploration of alternatives, 11–12
  limitations of use, 47

Decision-making power, waning of, 75–78
Delirium, 50–53
  associated conditions, 50–51
  case vignettes
    exam failure by teacher, 51–52
    from medication withdrawal, 52
  clinical features, 51t
  mortality data, 51
Delusional disorder
  DSM-IV-TR criteria, 135
  from severe depression, 37
Dementia, 53–54. See also Alzheimer's disease; frontotemporal dementias; minimal cognitive impairment (MCI)
  autopsy studies, 59
  brain changes (vascular), 57–58
  causes of, 58–59
    mixed, and prodromal period, 59–60
  depression and, 63
  early stage/diagnosable, 27
  hallmarks of, 53
  talking to patients/caregivers about, 160–161, 161t
Denial
  of caregivers, 72, 113
  by cognitively impaired individuals, 131, 141
Depakote® (valproate acid), 46
Dependent personality disorder, 17, 18, 47, 133
Depression. See also antidepressant medication
  additive demoralization from, 106
  antidepressant treatment, 28, 37
  case vignette, 16
  causes of, 38–43

alcohol, 43
general medical conditions, 40t–41t
medication, 41t, 42–43
and cognitive impairment, 47–48
in cognitively intact geriatric patients, 24, 25, 38–45
dementia and, 63
DSM-IV-TR criteria, 9t, 37, 38t, 46
ECT for, 37
IPT/medication combination for, 4–6, 23, 69
pharmacological causes, 41t–42t
three parts of, 9t
treatment for, 43–45
  IPT/PST, 44–45
  SSRIs, 43–44
talk therapy (See psychotherapy)
Diabetic retinopathy, 31
Diagnostic and Statistical Manual of Mental Disorders-IV
  delusional disorder criteria, 135
  depression criteria, 9t, 37, 38t, 46
Dialectical behavioral therapy (DBT), 44, 47
Dilantin® (phenytoin), 33
Donepezil (Aricept®), 55, 56, 58, 96, 119, 161, 170–171
Dopamine, 63
Driving cessation, strategies for presenting, 125, 125t–127t
Drowsiness during daytime, 34
Drug toxicity, 43
Dyadic role-dispute resolution, 25, 89

ECT. See electroconvulsive therapy (ECT)
Ego integrity vs. despair (last stage of life), 34–35
Elder abuse (case vignette), 111–113
Electroconvulsive therapy (ECT), 12, 37, 160, 169
Endocrine conditions and depression, 40t
Erickson, Erik, 34–35
Evidence-based treatment. See also cognitive-behavioral therapy (CBT)
  for depression, v, 4, 24–25, 44
  for obsessive compulsive disorder (OCD), 144
Executive dysfunction (ED), 27. See also cognitive impairment
  additive/disinhibiting role of alcohol, 70
  alcoholism and, 67, 150

Executive dysfunction (ED) (*continued*)
  causes of, 65–67
  driving safety and, 80
  Executive Interview for, 67
  gradual onset, recognition of, 79
  presentation of symptoms/case vignettes
    loss of decision-making power, 75–78
    sporadic odd behavior/aggression, 68–70
    spouse becoming caregiver, 70–74
  rating scales for, 67
  safety hazards due to (vignette), 81–83
  signs of, 65
  value of psychoeducation in, 80–83
Executive Interview (EXIT), 67
Exelon®, 55, 56, 161

Family rivalry and last wills and testaments, 171–173
Feil, Naomi, 23
Foci (problem areas) of IPT. *See* interpersonal deficit (interpersonal sensitivity); role disputes; role transitions; unresolved grieving
Follow-up, long term, 161–162
Freud, Sigmund, 21
Friends as caregivers, 155–156
Frontotemporal dementias, 53, 54, 70, 125
Future contingency planning, 166–171
  advanced directives, 36, 88, 113*t*, 173–174, 176
  last wills and testaments, 92*t*, 109, 166, 171–173, 176
  power of attorney, 29, 36, 92*t*, 166, 173

Generalized anxiety disorder (GAD), 46, 142, 166
Geriatric mental health specialists, 176–177
Geriatric patients. *See also* cognitively intact geriatric patients
  depression issues, 38–45
  effects of drugs in, 33
  housing options for, 30–31
  new, initial interviews with, 94–95
  role transition difficulties, 25
  sleep problems, 34
Gerontology/geriatric medicine
  background information
    aging process, 31
    psychologic changes of aging, 31–32

    effect of medications, 33
    effect of medications on elderly, 33
    polypharmacy/drug–drug interactions, 32–33
    sleep problems, 34
Grief/grieving process
  abnormal, in cognitive impairment, 121–124
  complicated (case vignette), 13–14
  coping difficulties, 18
  examples, 118*t*
  unresolved, 10, 11, 87, 100

Healthcare collaborative model, 28–29
Hearing loss, 31
Hormones and depression, 42*t*
Hydration of body (with aging), 32

Imaging studies (of the brain)
  CAT scans, 54, 62
  MRI scans, 54, 62, 81, 123
  neuropsychological testing, 62
  SPECT scans of, 54, 62, 70
Immunization programs, 30
IMPACT study, 29, 44, 48
Infections/inflammatory conditions and depression, 40*t*
Insight-oriented psychotherapy, 154
Insomnia, 34, 44
  psychotropic medications for, 29
Instrumental activities of daily living (IADL), 5, 26, 30, 35
Interpersonal deficit (interpersonal sensitivity)
  case vignettes
    declining function from cognitive impairment, 130–132
    dependent personality disorder, 17–20
    schizoid personality disorder, 136–137
  in patients with PD with cognitive impairment, 132–138
    antisocial personality disorder, 133
    borderline personality disorder, 134–135
    dependent personality disorder, 133
    narcissistic personality disorder, 133–134
    obsessive-compulsive personality disorder, 137–138
    paranoid personality disorder, 135–136

Index 195

schizoid/schizotypal personality disorder, 136–137
Interpersonal psychotherapy (IPT), traditional, 44
  antidepressant medication with, 4–6, 8, 23, 69, 75, 100, 160
  benefits, 11–12
  CBT/PST vs., 12
  for cognitive impairment, 24–25
  for cognitively intact geriatric patients (vignettes)
    adjusting to retirement, 14–15
    complicated grief, 13–14
    dependent personality disorder, 17–20
    role dispute during adjustment to retirement, 15–17
  description of, 44–45
  development of, 8–9
  goal of, 9*t*
  individual vs. group therapy, 11
  phases of (I–IV), 9–10
  preparations for IPT-ci, 20
  three parts of depression, 9*t*
Interview (in IPT-ci)
  of family members, 89*t*, 99, 159
  initial, of patient, 51, 88, 89, 94–98
  parallel, by coworker, 147
IPT-ci (interpersonal psychotherapy for use with cognitively impaired adults). *See also* therapists (of IPT-ci)
  delivery in common care settings, 27–28
  description, v–vi
  future directions for, 175–178
  future research needs, 177–178
  goals/strengths of, 4, 24–25
  implementation (who) decision, 28–29
  preparation for, 20
  rationale/background for, 21–29
  research using, 4–7
  as training paradigm, 176–177
IPT-ci basics
  contract formation, 100–102
  initial interview, 94–98
  patients given "sick role," 99–100
  psychoeducation in phase I, 9, 87, 89, 98–99, 159
  separate family member visits, 99
  17 basic steps (summary), 88*t*–92*t*
  strategies for caregivers, 93*t*–94*t*
  summary of, 175–176
  treatment conceptualization (three phases), 87–88
IPT-ci engagement variations, 158*t*, 159–160

Lamictal® (lamotrigine), 46
Last wills and testaments, 92*t*, 109, 166, 171–173, 176
Life care communities, 36
Lifespan, 30–31, 36
Lithium, 46
Longevity record, 30
Long-term planning, 164–165
Lorazepam, 46

Macular degeneration, 31
Magnetic resonance imaging (MRI) scans, 54, 62, 81, 123
Maintenance Therapies in Late-Life Depression Study I (MTLD-1), 4–5, 100
Maintenance Therapies in Late-Life Depression Study-II (MTLD-2), 5–6, 23
Malabsorption syndrome, 59
Manual-based psychotherapy. *See* interpersonal psychotherapy (IPT), traditional
Medical conditions with depressive features, 40*t*–41*t*
Medication. *See also* antidepressant medication; psychotropic medications
  acetylcholine, 55, 56, 58, 63
  alprazolam, 46, 52
  anticholinesterases, 41*t*
  Aricept® (donepezil), 55, 56, 58, 96, 119, 161, 170–171
  aripiprazole, 46
  benzodiazepines, 46
  Depakote® (valproate acid), 46
  Dilantin® (phenytoin), 33
  effects in the elderly, 33
  Exelon®, 55, 56, 161
  Lamictal® (lamotrigine), 46
  lithium, 46
  lorazepam, 46
  olanzapine, 46
  paroxetine, 5–7
  quetiapine, 46

Medication (*continued*)
  Razadyne®, 55, 56, 161
  sedative drugs, 34, 42*t*, 50, 79, 146, 147
  SSRIs, 43, 70, 73, 145
  valproic acid (Depakote®), 46
  Xanax® (alprazolam), 52
Memory deficits, 3, 62, 182
Memory-enhancing agents, 105, 130, 161, 167
Minimal cognitive impairment (MCI), vi, 3, 48, 60–61, 70
  amnestic type, 75, 118–119
  progression of
    to Alzheimer's disease, 167
    to frontotemporal dementia, 73–74
Mini-Mental Status Exam (MMSE), 4–5, 53
  administration at initial interview, 96–97
  executive function (vignette), 69
  limitation of, 60–61
Modeling by therapist
  of calming suggestions, 120
  for caregivers, 134–135
  of respectful behavior, 89, 94–95, 99
Modeling (by therapist) of respectful behavior, 89*t*, 94–95, 99
Montreal Cognitive Assessment (MOCA), 67, 68*f*, 97
  administration/scoring instructions, 179–182

Namenda®, 55, 56, 161
Narcissistic personality disorder, 46–47, 133–134
National Health Service (Great Britain), 56
National Institute for the Mentally Handicapped (NIMH), 4
Neoplastic conditions and depression, 40*t*
Neuroleptics (atypical), 46
Neurological agents and depression, 42*t*
Neuropsychological testing, 21, 28, 29, 62, 70, 73, 90*t*, 98, 118–119, 167, 170
Neuropsychologists, 29, 48, 61
Nonfamily caregivers, 116, 155–156
Norepinephrine, 63
Nurses/nurse practitioners, 12, 28–29, 30, 35, 48, 147
Nursing homes, 28

Obsessive-compulsive disorder (OCD), 46–47, 137–138, 142–144
Olanzapine, 46
Oppositional behavior, 29

Panic disorder, 46
Paramour caregivers, 156–157
Paranoid personality disorder, 135–136
Parkinson's disease, 24
Paroxetine, 5–7
Passivity, 17, 65, 109–110, 170
Personality disorders
  borderline personality disorder, 46–47, 134–135
  dependent personality disorder, 17, 18, 47, 133
  narcissistic personality disorder, 46–47, 133–134
  obsessive-compulsive disorder, 46–47, 137–138, 142–144
  paranoid personality disorder, 135–136
  schizoid/schizotypal personality disorder, 37, 110–111, 136–137
Phenytoin (Dilantin®), 33
Phobias, 46
Physician assistants, 28
Physiologic changes of aging, 31–32
Planning
  as concern of caregivers, 93
  family involvement, 110, 153
  for future contingencies, 166–171
  lack of, with ED, 65–66, 76
  for long-term, 164–166
Polypharmacy, 32–33
Posttraumatic stress disorder (PTSD), 46
Power of attorney, 29, 36, 92*t*, 166, 173
Primary care physicians, 28, 29
PRISM-E studies, 48
Problem-solving therapy (PST), 12, 44
Prodromal period of dementia symptoms, 59–60
PROSPECT study, 29, 44, 48
Prostate gland enlargement, 34
Psychiatrists, 28
Psychoeducation, 24–25, 74
  for caregivers, 147
    who are spouses, 113–115
  for families, 110
  patient benefit from, 78
  patient need for, 28, 59, 97
  in phase I of IPT, 9, 87, 89, 98–99, 159
  therapist preparation for, 20
  value of in ED, 80–83

Psychosis, 29, 37
Psychotherapy. *See also* cognitive-behavioral therapy (CBT); insight-oriented psychotherapy; interpersonal psychotherapy (IPT), traditional; problem-solving therapy (PST)
 antidepressant medication vs., 44
 caregiver need for, 91
 in collaborative care models, 48–49
 influence of memory impairment on, 21–23
 supportive, offered by social workers, 28
Psychotropic medications, 34, 87, 89*t*, 90*t*
 with CBT, 144
 compliance monitoring for, 160
 for decreasing impulsivity, 59
 for depression/dementia, 177
 fat solubility of, 42–43
 with IPT, 29, 87, 144, 163, 166–167
 for OCD, 143

Quetiapine, 46

Razadyne®, 55, 56, 161
Respite care, 28
Restless leg syndrome, 34
Retirement (role transition)
 adjusting to, 14–15
  role dispute during, 15–17
Retirement communities (as "brain trusts"), 35
Role disputes. *See also* dyadic role-dispute resolution
 communication analysis in, 10
 contribution to depression, 94
 examples of, 117*t*, 118–121
 IPT-ci therapist role, 87, 96, 100, 101
 between parent
  and adult child, 107, 127–129
  and other family members, 116
 resolution of, 25, 87, 89, 128, 163, 175
 of spouse caregiver with spouse, 149–151
Role disputes, case vignettes
 with alcohol-induced delirium, 140–142
 depression/waning cognitive ability, 168–171
 with husband over misuse of prescribed meditation, 144
 over embarrassing restaurant behavior, 149–151

 parent–adult child, 127–129
 son's child-rearing practices, 118–121
Role playing, 10, 16, 75, 100, 134, 163
Role transitions, 8, 10
 adjusting to retirement, 14–16
 executive dysfunction (vignette), 68–70
 role dispute during, 15–17
 of caregivers, 147–157
 of adult children, 107, 148
 of spouse, 148–151
 coping skills/strategies for, 87
 lifelong OCD with cognitive impairment, 142–144
 when cognitively impaired, 124–125
 elder abuse and, 111–113
 examples of, 117*t*
 grieving process, 121–122
 of identified patient/caregiver, 4
 need to stop driving, 125, 125*t*–127*t*
 in traditional IPT, 69, 75, 100

Sanitation improvements, 30
Schizoid/schizotypal personality disorder, 37, 110–111, 136–137
Sedative drugs, 34, 42*t*, 50, 79, 146, 147
Sedatives/hypnotics
 and delirium, 52
 and depression, 42*t*
Selective serotonin reuptake inhibitors (SSRIs), 43, 70, 73, 145
Sense diminishment (with age), 31
Siblings
 as caregiver, 103, 109
 depression vulnerability risks, 38–39
 mutual interdependence of, 154–155
 rivalry between, 155
 sibling-tension issue (vignette), 76–77
"Sick" role of patient, 9, 16, 87, 91, 99–100
Skin issues (aging), 32
Sleep problems, 34, 44, 46
Smoking, 33, 57, 58
Social support, importance of, 35
Social workers, 28–29, 48, 166
Somnambulism (sleepwalking), 34
SPECT scans, 54, 62, 70
Spouse caregivers, 113–115
 case vignettes, 70–74, 149–154
 cognitive decline of, 102
 marital strain issues, 115

Spouse caregivers (*continued*)
   psychoeducation of, 113–115
   relationship questions, 113*t*
   role transitions for, 148–151
   soliciting viewpoint of, 114
Steady state
   described, 92*t*, 162
   as goal of IPT-ci treatment, 87, 91–92, 102
   long-term planning and, 164–166
   maintenance of, 163–164
Steroids/hormones and depression, 42*t*
Stimulants and depression, 42*t*
Stress of caregiver, 148*t*
Suicidal ideation, 49

Talk therapy (psychotherapy). *See* psychotherapy
Therapists (of IPT-ci)
   addiction strategy, 138
   advocacy for patients by, 171, 173–174
   borderline personality disorder strategy, 134–135
   brainstorming with family, 97–98
   caregiver work, 20, 26, 101, 109–110, 116, 135, 147, 150, 163
   collaboration with PST therapist, 44
   encouragement role of, 72, 77
   establishment of patient connections, 23, 96
   familiarization with local laws, 166
   follow-up/monitoring of patients, 78, 161–162, 175
   intermediary role of, 73, 76, 78, 107, 114, 128, 130, 157
   issues presentation by, 88
   modeling by, 89, 94–95, 99, 120, 134–135
   patient assessment by, 95, 96, 99, 160
   patient-related goals of, 27
   planning for future contingencies, 166–171
   psychoeducation role of, 24–25, 74, 91, 119
   role of, in 17 steps of IPT-ci, 88*t*–92*t*
   role transition focus by, 10, 14, 168–171
   schizotypal PD strategy, 136
   structure of therapy, 11
   talking to patients/caregivers about dementia, 160–161, 161*t*
Thyroid hormone, 61
Tobacco use, 33, 57, 58
Traditional interpersonal psychotherapy (IPT). *See* interpersonal psychotherapy (IPT), traditional

Unresolved grieving, 10, 11, 87, 100

Validation Therapy (Fell), 23
Valproic acid (Depakote®), 46
Vascular dementia, 62
Vision issues (of aging), 31
Visual-spatial difficulties, 70
Vitamin B1 (thiamine) deficiency, 61
Vitamin B12 deficiency, 59

Water purification, 30
Wernicke–Korsakoff syndrome, 61
Wills and testaments, 171–173
Withdrawal issues
   from alcohol, 53, 140
   from sedative drugs, 50, 52, 53
   from stimulants, 33

Xanax® (alprazolam), 52